AF748812

Literary Criticism and Cultural Theory

Edited by
William E. Cain
Professor of English
Wellesley College

A Routledge Series

Literary Criticism and Cultural Theory

William E. Cain, *General Editor*

An Ethics of Becoming
Configurations of Feminine Subjectivity in Jane Austen, Charlotte Brontë, and George Eliot
Sonjeong Cho

Narrative Desire and Historical Reparations
A.S. Byatt, Ian McEwan, Salman Rushdie
Tim S. Gauthier

Nihilism and the Sublime Postmodern
The (Hi)Story of a Difficult Relationship from Romanticism to Postmodernism
Will Slocombe

Depression Glass
Documentary Photography and the Medium of the Camera Eye in Charles Reznikoff, George Oppen, and William Carlos Williams
Monique Claire Vescia

Fatal News
Reading and Information Overload in Early Eighteenth-Century Literature
Katherine E. Ellison

Negotiating Copyright
Authorship and the Discourse of Literary Property Rights in Nineteenth-Century America
Martin T. Buinicki

"Foreign Bodies"
Trauma, Corporeality, and Textuality in Contemporary American Culture
Laura Di Prete

Overheard Voices
Address and Subjectivity in Postmodern American Poetry
Ann Keniston

Museum Mediations
Reframing Ekphrasis in Contemporary American Poetry
Barbara K. Fischer

The Politics of Melancholy from Spenser to Milton
Adam H. Kitzes

Urban Revelations
Images of Ruin in the American City, 1790–1860
Donald J. McNutt

Postmodernism and Its Others
The Fiction of Ishmael Reed, Kathy Acker, and Don DeLillo
Jeffrey Ebbesen

Different Dispatches
Journalism in American Modernist Prose
David T. Humphries

Divergent Visions, Contested Spaces
The Early United States through the Lens of Travel
Jeffrey Hotz

"Like Parchment in the Fire"
Literature and Radicalism in the English Civil War
Prasanta Chakravarty

Between the Angle and the Curve
Mapping Gender, Race, Space, and Identity in Willa Cather and Toni Morrison
Danielle Russell

Rhizosphere
Gilles Deleuze and the "Minor" American Writings of William James, W.E.B. Du Bois, Gertrude Stein, Jean Toomer, and William Faulkner
Mary F. Zamberlin

Rhizosphere

Gilles Deleuze and the "Minor" American Writings of William James, W.E.B. Du Bois, Gertrude Stein, Jean Toomer, and William Faulkner

Mary F. Zamberlin

Routledge
New York & London

Excerpts from *A Thousand Plateaus: Capitalism and Schizophrenia*, by Gilles Deleuze and Félix Guattari, translated by Brian Massumi, used by permission of the University of Minnesota Press.

Excerpts from *Mille Plateaux Capitalisme et Schizophrénie*, by Gilles Deleuze and Félix Guattari, copyright 1980 by Les Editions de Minuit, reprinted by permission of George Borchhardt, Inc. for Les Editions de Minuit.

Excerpts from *Cane*, by Jean Toomer, reprinted by permission of Liveright Press.

Excerpts from *The Souls of Black Folk* by W.E.B. Du Bois, copyright 1989 by Viking Penguin, used by permission of Penguin, a division of Penguin Group.

Published in 2006 by
Routledge
Taylor & Francis Group
270 Madison Avenue
New York, NY 10016

Published in Great Britain by
Routledge
Taylor & Francis Group
2 Park Square
Milton Park, Abingdon
Oxon OX14 4RN

© 2006 by Taylor & Francis Group, LLC
Routledge is an imprint of Taylor & Francis Group

Printed in the United States of America on acid-free paper
10 9 8 7 6 5 4 3 2 1

International Standard Book Number-10: 0-415-97535-2 (Hardcover)
International Standard Book Number-13: 978-0-415-97535-3 (Hardcover)

No part of this book may be reprinted, reproduced, transmitted, or utilized in any form by any electronic, mechanical, or other means, now known or hereafter invented, including photocopying, microfilming, and recording, or in any information storage or retrieval system, without written permission from the publishers.

Trademark Notice: Product or corporate names may be trademarks or registered trademarks, and are used only for identification and explanation without intent to infringe.

Library of Congress Cataloging-in-Publication Data

Catalog record is available from the Library of Congress

informa
Taylor & Francis Group
is the Academic Division of Informa plc.

Visit the Taylor & Francis Web site at
http://www.taylorandfrancis.com

and the Routledge Web site at
http://www.routledge-ny.com

In Loving Memory: I dedicate this book to my father Anthony Dominic Zamberlin and brother Mark Philip Zamberlin

Contents

Acknowledgments

I wish to express gratitude to Professor Mikkel Borch-Jacobsen for encouraging my work. I also thank my daughter Marguerite Claire for her luminous presence and my mother Colleen.

Introduction

Between Gilles Deleuze and William James: Rhizomatics and Pragmatics

"Yes, we will all have loved philosophy. Who can deny it? But, it's true, (he said it), Deleuze was, of all those in his "generation," the one who "did/made" (faisait) it the most gaily, the most innocently. . . . This was no doubt the condition for his having left a profound mark on the philosophy of this century, the mark that will remain his own, incomparable. The mark of a great philosopher and a great professor. This historian of philosophy who proceeded with a sort of configurational election of his own genealogy . . . was also an inventor of philosophy who never shut himself up in some philosophical "realm." . . . And then and then I want to say precisely here that I loved and admired his way—always faultless. . . . I would have tried to tell him why his thought has never left me, for nearly forty years. How could it ever do so?"

—Jacques Derrida

"It is not by means of an exegetical practice that one could hope to keep alive the thought of a great thinker who has passed away. Rather, such a thought can only be kept alive through its renewal, by putting it back into action, reopening its questioning, and by preserving its distinct uncertainties- with all the risks that this entails for those who make the attempt."

—Félix Guattari

February seventh, in the year 2000 on the University of Washington campus, the French novelist and poet Maurice Dantec speaks of his chosen exile to North America in order to write truly modern *American* works. In his discourse it becomes evident that *American* literature need not necessarily originate from a North American citizen, but rather certain *modern American* qualities would distinguish it as such, distinguish it from the art of an old and “dead” world, that of his homeland France. To questions on his reading and description of his work Dantec responded with an unmistakably Deleuzian artillery of terms such as, deterritoritalize, nomadism, the marking out of new stylistic terrains. Dantec as does Gilles Deleuze, juxtaposes American Literature to that of France to elevate it via the contrasts. To paraphrase Dantec, American modern literature vitalizes a lapsing present towards a future movement. Conversely France, a country that views itself as the guardian of the “civilized world’s” fading traces or decomposing relics produces aesthetic products steeped in past-bound cultural, historical values, signs and points of departure to nullify any sense of an immediate, vital, present energy that holds a future potential. When pursued by questions from French literature scholars on certain passages in his novel that unmistakably betray traces of the French obsession with the “other,” Dantec reluctantly admits that certain remainders of his French literary, educational, cultural “baggage” can never be completely dropped off. Despite this Dantec all the more emphatically summarizes his major points with the remark: “Le base du roman moderne est en Amérique.” (*The base of the modern novel is in America.)*

Maurice Dantec’s positive description of American literature echoes remarks made by various French intellectuals throughout the twentieth century and as already noted Dantec evidently finds much of the fuel for his fire in the writings of Gilles Deleuze, who writes in *Mille Plateaux:*

> “Nevertheless, everything important that has happened or is happening takes the route of the American rhizome: American books are different from European books, even when the American sets off in pursuit of trees. The conception of the book is different. *Leaves of Grass*. . . . Make rhizomes, not roots, never plant! Don’t sow, grow offshoots! Don’t be one or multiple, be multiplicities! Run the lines, never plot a point! Speed turns the point into a line!. . . . As they say about old man river: *He don’t plant ‘tatos/ Don’t plant cotton/ Them that plants them is soon forgotten/ But old man river he just keeps rollin’ along.”* (Mille Plateaux, 19, 24–5).

Evidently Dantec read much Deleuze and admired him, as he employs Deleuzian vocabulary terms to express his literary agenda. Moreover, Dantec continues to act as a disciple of Deleuze which his interpretation and improvisation of the latter’s thought in his collaborative efforts with one of Deleuze’s colleagues and friends, the writer and musician Richard Pinhas make manifest. (Pinhas also directs the Deleuze web site.) These two philosophers/artists have given Deleuze inspired concerts throughout Europe and the United States and have produced a recording entitled *Schizotrope,* which puts Deleuze's word to abstract techno-music sounds. The homage these aesthetic/philosophical products render to Deleuze's thought result in a sign that speaks to Deleuze's demarche that blurs the line that separates art and philosophy. In fact In *Dialogues,* Deleuze states that instead of philosophy being some sort of ultimate discipline, creations generated in *fields* outside of philosophy, in primarily aesthetic realms, trace the lines that allow for a philosophical revelation, thought, or experience.

> "There is really no need for philosophy: It is necessarily produced there where each activity extends philosophy's lines of deterritorialization. Get out of philosophy, do anything in order to produce it from the outside. Philosophers have always been other things, they are born of other things."

> « Il n'y a aucun besoin de philosophie : elle est forcement produite là où chaque activité fait pousser sa ligne de déterritorialisation. Sortir de la philosophie, faire n'importe quoi, pour pouvoir la produire du dehors. Les philosophes ont toujours été autre chose, ils sont nés d'autre chose. » (Dialogues, 89)

One group that Deleuze identifies as those "philosophers born from/of other things" that "get out of philosophy," and produce it from the outside, are American "rhizomatic" writers. Deleuze asserts that the reason philosophy as a "specialized institution" has grown almost extinct in America lies in the fact that American writers knew how to make a "pragmatics" out of their literature: knew who to move "between" things with the "logic of the AND." Due to this *know how* American "minor" literature overthrows ontological foundations, as its writing transmits thought as action, nullifies a sense of beginning and end and effaces distinctions between all subject object relations. "The grass and the road grow one into the other, the becoming-buffalo." (Mille Plateaux 37, Dialogues 38)

The preceding paragraph with its *indefinite* metaphors or *deleuzian* "concepts" such as "rhizomatic," "devenir-"(becomings), may not be readily deciphered applying a traditional philosophical canon of reference terms but Deleuze uses them expressly to evoke multi-dimensional meanings rather than provide one readily comestible idea that holds little meaning. Other such terms if taken out of a Deleuzian context, including the signifiers, *minor* and *major* literature, *livre-rhizome and livre-racine, minoritaire,* and *majoritaire, Molar* or *arborescent* systems, lignes de fuite (*lines of flight),* and the list goes on, resist being "explicated" in a field of isolation. Therefore this writing hopes that with the patient participation of the reader these "concepts" and their meanings will more effectively be gleaned and grasped in a context of questioning and application. Otherwise the struggle to "understand" or "explain" too precipitously will only negate these meanings in the same gesture. One will however attempt to provide the reader with a certain sense of the term "rhizome" and its adjacent signifiers such as "rhizomatic," "livre-rhizome" (book-rhizome), "rhizosphere," to help give an indication of the directions and spirit of this writing. For Deleuze, "rhizome" as a metaphor may be applied to anything that generates life, heterogeneous connections and mutant lines. As already cited on page two, Deleuze tells us that "minor" American books make "rhizome," off-shoots, multiplicities; they make/do a pragmatics. In Deleuze's writings these two expressions, making rhizome or a pragmatics, become interchangeable. In the section of *Mille Plateaux* entitled "Rhizome" Deleuze writes:

> "Schizo-analysis or pragmatics have no other sense: make/do rhizome. . . . The rhizome in itself has very diverse forms, from ramified surface extension in all directions to concretion into bulbs and tubers . . . we will convince no one unless we enumerate certain approximate characteristics of the rhizome. 1 and 2. Principles of connection and heterogeneity: any part of a rhizome can be connected to anything other and must be. This is very different from the tree or the root, which plots a point, fixes an order." (Thousand Plateaux, 7)

> « La schizo-analyse ou la pragmatique n'ont pas d'autre sens: faites rhizome. . . . Le rhizome en lui-même a des formes très diverses, depuis son extension superficielle ramifiée en tous sens jusqu'à ses concrétions en bulbes et tubercules . . . nous ne convaincrons personne si nous n'énumérons pas certains caractères approximatifs du rhizome. 1 et 2. Principes de connexion et d'hétérogénéité : n'importe quel point d'un rhizome peut être connecté avec n'importe quel autre, et doit l'être. C'est très différent de l'arbre ou de la racine qui fixent un point, un ordre. » (Mille Plateaux, 13)

Literally, the term "rhizome" refers to a certain plant life, (bulbs, tubers, weeds), *The Cambridge International Dictionary of English* (p.1220), defines rhizome as, "A stem of some plants which grows horizontally along or under the ground." *Braken* is a type of fern which often grows in woods and hearths where it may cover extensive areas by means of its rhizomes." According to the *Oxford English Dictionary, phanerogams,* which spring from rhizomes organically represent both union, fertilization and fecundity: The definition notes the Greek etymology: Greek phanero oaverpo-s,= visible, evident, + yauos marriage, sexual union. Although "rhizome" literally refers to plant life, Deleuze also represents it as a musical writing in the opening chapter to *Mille Plateaux,* entitled "Rhizome." This writing is not a typical musical composition with recognizable notes generating familiar sounds. Rather the notes have turned into lines that transgress the boundaries separating the five musical bars and swirl and venture from one bar to the next becoming intertwined to result in both tangled knots and straight extensions. Finally the composition resembles a map whose transformed and transformative notes, sites and sounds all move on in a plurality of directions despite the normative, surface rules that would have the notes remain distinct and distinguishable points positioned on one particular line to generate one specific, recognizable sound.

In a "rhizomatic" operation whether, organic, musical, writing or other, an infinite variety of heterogeneous lines intersect, pick up lines left off, co-penetrate and mutate through co-*agencements,* that carry and proliferate a multitude of particles from an innumerable array of semiotic systems, to co-create infinite, deterritorialized, a-centered, a-systematized "worlds." In this map making, or cartographic, nomadic movement, there are no pre-established paths, or destinations, based on pre-existent notions and ideologies or mediated "major order" desires. One can say that "rhizome," as *musical form,* or more precisely that musical forms, or writings that operate as "rhizomes," *in their ruptures and proliferations,* become hyper fecund, and multiply "lives" as they "trace virtual lines of infinite variation." "While putting all components in continual variation, music itself becomes a sub-linear system, a rhizome instead of a tree, and serves a cosmic virtual continuum, wherein even the gaps, silences, ruptures, cuts play a part. . . . Music hasn't ceased to circulate its lines of flight, as well as so many "multiplicities of transformation," even in reversing its proper codes that structure or "arbrifient" it; this is why the musical form, in its ruptures and proliferations, is comparable to the weed, a rhizome."(Mille Plateaux, 121, 19). All of these words or images used to "approximate" the prolific, creative, life giving forces of "rhizome" speak to its unpredictable movements, mutations and metamorphoses as well as its irreducible qualities and nature. As far as the term's application to art and philosophy, philosophy that makes "rhizome," co-creates "vitalistically," as it thinks *with,* writes *with,* creates *with* a plethora of particles from diverse sign regimes to provide for new languages, and modes of thought, instead of reducing, and extinguishing life forces through ontological, interpretive gestures that recognize, judge, and plot points.

> "One writes in relation to a people to come, who have not yet a language. To create is not to communicate, but to resist. . . . This is the power of inorganic life, that which can take place in the line of drawing, writing, or music. Organisms die, not life. There is not an oeuvre that indicates a dead end to life, one that doesn't trace out a pathway between the stones. All that I wrote was vitalist, at least I hope."

> « On écrit en fonction d'un peuple à venir et qui n'a pas encore de langage. Créer n'est pas communiquer, mais résister. . . . C'est la puissance d'une vie non organique, celle qu'il peut y avoir dans une ligne de dessin, d'écriture ou de musique. Ce sont les organismes qui meurent, pas la vie. Il n'y a pas d'œuvre qui n'indique une issue à la vie, qui ne trace un chemin entre les pavés. Tout ce que j'ai écrit était vitaliste, du moins je l'espère. » (Pourparlers, 196)

Art that makes rhizome, (a "pragmatics"), generates an *Outside*-philosophy philosophy, one of multiplicities, that articulates new enunciations, other desires, and produces the unconscious itself; one free of the "dictatorial powers of psychoanalysis," its theories and methods, and those of the "Major" world order that submit the unconscious to *arborescent* structures, hierarchical graphs, recapitulating memories and the central organs, the phallus, the "arbre-phallus" (tree-phallus). (Mille Plateaux, 27).

> "In treating the unconscious as an a-centered system, this is to say as a mechanic network of finite automata (rhizome), the schizo-analysis, attains a completely different state of the unconscious. . . . For both enunciations and for desires, the question is never to reduce the unconscious, to interpret it or make it signify following the tree. The question is to produce the unconscious, and, with it, new enunciations, and other desires: the rhizome is this production of the unconscious itself."

> « En traitant l'inconscient comme un système acentré, c'est à dire comme un réseau machinique d'automates finis (rhizome), la schizo-analyse atteint à un tout autre état de l'inconscient. . . . Pour les énoncés comme pour les désirs, la question n'est jamais de réduire l'inconscient, de l'interpréter ni de le faire signifier suivant un arbre. La question, c'est

de produire de l'inconscient, et, avec lui, de nouveaux énoncés, d'autres désirs : le rhizome est cette production d'inconscient même. » (Mille Plateaux, 27).

This text that titles itself "*Rhizosphere,*" hopes to allow for an ambiance, plan(e), cartography, or fluid space, that allows the lines, of both the "minor"/ "rhizomatic"/ "pragmatic" aesthetic and philosophical writings that it stems from, (that traditional interpretive gestures hold apart as very different operations), to intersect, interpenetrate, inter-engage, until the normal, or "molar" lines that separate them blur and fade. This writer deliberately chooses four different American literary texts as it believes them to manifest the "pragmatic" way of the rhizome particularly well. These writings like those of Woolf, Proust, and Kafka, generate excessive, inclusive, overflowing "plans of immanence," that blend diverse lines and intersect with a multitude of sign regimes, to result in "minority machines" that activate "minoritarian becomings."[1] "Not a plan in the mind/spirit, but a real, immanent plan, non pre-existent, which blends all the lines, the intersection of all regimes (composing diagrammatic): Virginia Woolf's Wave, Lovecraft's Hypersphere, Poust's Spinder's Web, Kleist's Program, Kafka's K-Function, the Rhizosphere." (Dialogues II, 122, 123). Deleuze tells us that in the "rhizoshpere" fixed distinctions between content and expression are extinguished. Only particles remain that interact and enter into even supposedly opposed proximities to generate an immanent plan. In this plan there are no longer forms, subjects or persons organized in terms of a structure, or developed in terms of a genesis. Due to these factors that encourage and allow for innovative creations, rhizospheric writings act as "minority machines."

In this "rhizosphere," the mutant literary lines of four different American writers being: W.E.B. Du Bois, Gertrude Stein, Jean Toomer and William Faulkner not only co-connect and inform one the other but also intersect and engage with the philosophical writings of names such as, Gilles Deleuze, Félix Guattari, Jean Wahl, Jean Paul Sartre, Friedrich Nietzsche, and William James. All of theses lines, together called up and cited, proceed to move between, next to, into, over, upon one the other to intertwine and generate a space between art and philosophy, a mi-lieu where both co-function, co-produce, co-signify and co-create. This space, or plan(e) of consistence, hopes to operate in a manner that indicates a sense of "rhizome," not as "representation" as something sterile and stopped, but rather as a plan in process, perpetuating an active movement, a shifting, impression, that provides, glimpses and intuitions into its "resistant" life forces.

> "A rhizome neither begins nor ends, it is always in the middle, between things, inter-being, intermezzo. The tree is filiation, but the rhizome is alliance, uniquely alliance. The tree imposes the verb "to be," but the rhizome has for its tissue, the conjunction "and . . . and . . . and." There is enough force in this conjunction to shake up and un-root the verb "to be." . . . Between things does not designate a localizable relation that goes from one to the other reciprocally, but a perpendicular direction, a transversal movement which carries one and the other with it, the stream without beginning or end, which gnaws at its banks and picks up speed in the middle."

> « Un rhizome ne commence et n'aboutit pas, il est toujours au milieu, entre les choses, inter-être, intermezzo. L'arbre est filiation, mais le rhizome est alliance, uniquement d'alliance. L'arbre impose le verbe « être », mais le rhizome a pour tissu la conjonction « et . . . et . . . et ». Il y a dans cette conjonction assez de force pour secouer et déraciner le verbe être. . . . Entre les choses ne désigne pas une relation localisable qui va de l'une à l'autre et réciproquement, mais une direction perpendiculaire, un mouvement transversal qui les emporte l'une et l'autre, ruisseau sans début ni fin, qui ronge ses deux rives et prend vitesse au milieu. » (Mille Plateaux, 36, 37)

In relation to the rhizome as a middle space of movement that combines elements from opposite banks, that proliferates lines of flight that shoot off making more "stable" and "stabilizing" structures take off at the same time, this reading/ writing must be regarded as a double gesture. Herein, rather than explicating Deleuze's philosophical ideas, or "interpreting" the "rhizomatic," "pragmatic," American, literary text's selected, it tries to exemplify what Deleuze means by an active, interactive, plan(e), made of a multiplicity of various genres, strains of thought and impersonal names or "sign regimes." In other words, this text *reads with* Deleuze to help bring out the literary texts "micro-political" "lines of flight," and immaterial signs among other qualities. In the same spirit while one reads with other *pragmatic* philosophers and fiction artists one hopefully illuminates some of the loves, life, politics, creative ideas and inspirations that allowed Deleuze to write, think, philosophize "the most gaily" without guilt, to invent and produce differences with positive passion rather than simply reduce and replicate standard-ized,-izing readings and representations. As Deleuze insists, to create non-preexisting worlds, one needs to write and speak with the world, its elements and *persons.* Not to converse but rather to *conspire.* He refers to this activity as "agencing"; *being in the middle* to trace a line that includes both the interior and exterior world. Deleuze refers to this "with" movement as one that involves both body and "soul": contrary to a religious sense that dichotomizes these terms to elevate the soul above the body, Deleuze's soul intermingles and moves "with" bodies, with *their flesh* to extend innumerable life lines.

> "*All the subtle sympathies of the infinite soul.* . . . The soul is neither above or inside, it is "with," it is on the road, exposed to contacts, encounters . . . "feel with them, seize the vibration of their soul and their flesh in the passage," it is the contrary to a moral of salutation, teach the soul to live its life rather than to save it." (Dialogues 66, 77)

> « *Toutes les subtiles sympathies de l'âme innombrable* . . . l'Ame et le Corps, l' âme n'est ni au-dessus ni au-dedans, elle est « avec, » elle est sur la route, exposée à tous les contacts, les rencontres . . . « sentir avec eux, saisir la vibration de leur âme et de leur chair au passage », le contraire d'une morale de salut, enseigner à l' âme à vivre sa vie, non pas à la sauver. » (Dialogues, 66, 77)

To "write" in this Deleuzean "with" way that remains true to what Deleuze names a *pragmatic* or *rhizomatic* process is indeed a "different" and perhaps audacious endeavor. But only an "experimentation" promoted in this spirit, can potentially shed a degree of light on the "micro-political" implications of both "pragmatic," philosophical and aesthetic readings and writings whose lines and "lives" cross over and inseminate each (into) the other in a pro-genitive process. As this writing advances, the extent to which, "outside," "minor," "rhizomatic," or "pragmatic" aesthetic products provide potent, "philosophical," immaterial materials, and the degree to which, the kind of "pragmatic" philosophy Deleuze calls for extends and accentuates the force, power, mutant "lignes de fuite"(lines of flight), the "arms" of these art forms, will hopefully become clear. This reading/writing will claim success if it generates, activates, illuminates, and perpetuates the movement of any of the mutant lines of this textual terrain.

Issues linked to the fact that Deleuze, qualifies his "kind of philosophy" as a "pragmatics," and also describes *rhizomatic* American texts as ones that "make a pragmatics," an *Outside-philosophy* philosophy, will also be discussed at length in this text.[2] The first chapter of this text moves closer to understanding what "pragmatic" and its synonym term "micro-politics" mean in a Deleuzian context as it traces certain inspirational lines that lead

Deleuze to an interest in "radical empiricism," and a pluralistic or a *multiplicitous* pragmatics. One will consider Deleuze's relationship to Jean Wahl and Jean-Paul Sartre, two thinkers, mentors, and "teachers" who play paramount roles in inspiring an interest in American literature and thought. One can say that his intersection with these two intellectuals along with other influences, leads to Deleuze's distinction between a major and minor literature, micro and macro political regimes and systems, and his writings that begin to fervently promote a *pragmatic,* creative philosophical process. "You see the importance for me to define philosophy according to the invention or the creation of concepts, that means being neither contemplative or reflexive or communicative, etc., but as a creative activity." (Variations, 4).

Deleuze's philosophy like American "minor" literature also overthrows ontology, gets rid of beginning, end and origins, a sense of fixity and fixation, to proliferate a sense of the "multiplities" that create, play, and operate in the "intermezzo" the in-between spaces that engender the "rhizomatic" terrain that makes philosophy's system not simply "*a perpetual heterogeneity*" but one that becomes a "*heterogenesis.*" (*Variations,* 4). The "concepts" that Deleuze generates out of creative activity expressively resist "interpretive" gestures that attempt to explain them in terms of structures, genealogical reference points, or ontological concepts. "Deleuze's philosophy does not willingly lend itself to the exercise of commentary. In affect the multiplicities that it spreads out from book to book resist genealogical or structural interpretations. It is their indefinitely variable contour that draws a twisted line whose tours of detours trace the outline of a surface of multiple dimensions." (Variations, 12).

In light of the fact that pragmatic literary and philosophical texts resist normative readings, one also writes in consideration of Deleuze's call for a "well mannered" kind of reading that responds to pragmatic creations such as *minor* or *rhizomatic* literature in a way that becomes participative rather than reductive: This way involves becoming *sensorially* attuned to and immersed in the operations of the text until one produces with and through them to drive them on. This kind of reading rather than interpreting involves listening to a text as one receives a song. Understanding or not is no longer an issue, concepts are like sounds, colors or intensities that either work for you or don't, that pass or don't pass. (*Qui vous conviennent ou non, Qui passent ou ne passent pas*). (Dialogues, 10)

Perhaps the first chapter of this writing that traces out certain lines of influence appears to contradict all that has been just said on "interpreting" Deleuze. However the associations made between Deleuze and other thinkers do not in any manner reduce Deleuze's thought to theirs. Due to the inherent "pluralistic" quality of rhizomatic thought, the intersections, the "in-between" experiences that occur with other thinkers result in "differences" that always already include traces from a multiplicity of other proper "names" or *regimes of signs.* In other words, whatever evolves out of the brief co-existence of different thoughts can never be reduced to any "two" names or any definite number of names because each thought always carries traces of a multiplicity of other names, and ideas. This text inquires into Deleuze's appreciation for these individuals above all because he singles them out as the "most important thinkers" in his entourage. To inquire into the intellectual projects of these men simply sheds light on a few of Deleuze's *differences,* to better grasp their implications and possible meanings. Again one hopes to follow, consider and write lines in such a way that remains in synch with the *Pragmatic, rhizomatic,* process that considers the immanent, entangled, complex lines that move between.

> "That which we call diverse names- . . . micro-politics, pragmatics . . . rhizomatic, cartography- has no other object or goal than the study of these lines, in groups or individuals."

> « Ce que nous appelons de nom divers- . . . micro-politique, pragmatique . . . rhizomatique, cartographie- n'a pas d'autre objet que l'étude de ces lignes, dans des groupes ou des individus. » (Dialogues, 152, 153)

For now it will suffice to briefly mention these two influences being Jean Paul Sartre and Jean Wahl. However one makes the case that what derives from or evolves out of Deleuze's encounter with Sartre's thought becomes something *other.* While Deleuze often deals with similar themes and problems, he articulates them in very different ways until they mediate very different issues and ideological directions. On the other hand, one might characterize the thought that develops out of intellectual exchanges with the second figure, Jean Wahl, as taking the form of "variations." The *Littré, Dictionnaire de la langue française,* defines "variation": "Musical term. Changes made in a tune through the addition of ornaments which nevertheless allow the basic melody and movement to be maintained. -SYN. VARICATION, CHANGE: Variation, consists in being sometimes one way, sometimes another; change consists only in ceasing to be the same. A change can be complete; variation allows much similarity to remain."

This text considers Jean Wahl as one of Deleuze's, "*keys in the wind" that enables Deleuze's mind to take off*" (Dialogues, 90).[3] Jean Wahl may also be described as a key figure that has been missing in or has been overlooked or forgotten by recent intellectual histories. This writing tries to convey the principle role the philosopher and poet Jean Wahl played in opening certain doors of thought that encouraged Deleuze's "anarchic" philosophical procedures: Wahl played a part in initiating Deleuze's interests in thinkers such as Nietzsche, Bergson, Kafka, and American "modern" literary works, and ideas that have to do with the "pluralist-pragmatism" of American thinkers such as William James.

Despite the paramount role that Jean Wahl played, in augmenting an interest in American literature and thought very early on in twentieth century France; in inspiring enquiries into American, philosophy, language, style, cultural phenomenon in thinkers such as Gilles Deleuze, and Jean-Paul Sartre amongst others, academic texts seldom mention Jean Wahl. Jean-Philippe Mathy's book *Extreme Occident* bears witness to Wahl's textual "disappearance." This extensive historical study spans the period from post World War I to our present time and considers the French intelligentsia's various responses and reception of American culture, literature, life and politics, but doesn't even mention Jean Wahl once in its three hundred plus pages, not even in a note, or in its bibliography. Mathy brings up the recent phenomena of a "pragmatization" of French thought in the work of *post* post-modern theorists but fails to mention Wahl, who Deleuze indicates as the major figure who exposed the French intellectual scene to American pluralism and pragmatism. This exposure and introduction began as early as 1920 with Wahl's publication, *Les Philosophes Pluralistes d'Angleterre et d'Amérique.* Mathy cites Richard Rorty who writes about the "new," intercontinental, intellectual convergence.

> "Richard Rorty . . . contends that the international division of philosophical labor between the Anglo-American and Continental traditions is presently being undermined by what he calls "the pragmatization of analytic philosophy." . . . One need only think of Michel Foucault's "tool boxes," a metaphor he borrows from the pragmatists. Gilles Deleuze, as we saw, is very fond of Anglo-American literature, whose superiority over the French literary tradition lies for him in its use of writing as experimentation rather than interpretation . . . or Jean Baudrillard's celebration of American society as a "utopia achieved," of Americans as people who "build the real out of ideas," and Europeans as those who "transform the real into ideas, or ideologies" . . . something like a pramatization of European thought, which has traditionally been dominated since the Enlightenment by rationalist and universalist models, is indeed taking place."(Extreme, 248, 257–8)[4]

Mathy speculates on this "pragmatization" at the book's epilogue without any sort of interrogation into what incited this '*new*' intellectual overlap. After his text's lengthy historical deliberation, his omission if unintentional nonetheless conveys a sense that this *pragmatization* appeared post sixties without any prior French movements that encouraged, informed or inspired this interest in pragmatic, pluralist thought. *Extreme Occident,* not unlike other scholarly texts that comment on Deleuze's American tastes, either omits information pertaining to the influence of Jean Wahl, or betrays an unawareness of the transference of interests that one can show passed between the latter as mentor/ professor, and his student Gilles Deleuze.

This text states with Deleuze, that Jean Wahl initiates a "counter-cultural" French appreciation for American ideas and art. One qualifies this as *counter cultural* for as Mathy's book documents; historically French intellectuals for the most part met American "things" with hostility, defensiveness and a degree of fear.[5] This text inquires into the role Jean Wahl played, as Deleuze indicates, in introducing his students, and France to not only, "American pluralist pragmatism," particularly that of William James but also to a vision of a *becoming-revolutionary* of philosophy in general. The aspects of "American" thought and literature that Wahl values also animate this particular discussion in Deleuzian terms. For Wahl's introduction of American literature and pluralist pragmatism emphasized the *democratic* dimensions and ideals of both *genres* that affirm a fluidity of irreducible, ever colliding, intermixing variables that perpetuate a malleable non-formed, formation of process, flux and fruition.

Two books in particular that Jean Wahl successively wrote and edited in the midst of the First and Second World War underline the democratic values inherent in American literature and William James's pluralist pragmatism. The second book especially articulates a sense of gratitude and appreciation for America's solidarity in the war effort and praises its literature that bears witness to the power of art to bridge ideological and cultural gaps. Wahl writes,

> "*Here is this* envoy from North America to North Africa, and through it to France, this sign of friendship in this little trembling day. The novelists are here: Hemingway, Faulkner, Caldwell, Steinbeck, and Henry Miller. We also wanted to see Dos Passos. That he is present in our thoughts. No other country has gone as far as America in the bouleversement, the upsetting, overthrowing, violent agitation, of the techniques of the novel, which are so characteristic of the twentieth century. . . . That this collection consecrated by France to America, by America to France, which appears on French soil, stand as the sign of an immense intercontinental

continuity. To make the world today requires nothing short of the world." (Ecrivains et Poètes, 5, 6).

This text *Les Ecrivains et Poètes des Etats-Unis d'Amérique* also includes works by *minority* writers such as Langston Hughes and Gertrude Stein, which substantiates the democratic principles that it praises. This book produced in the onslaught of World War II symbolically and culturally desegregates and protests levels of psychological, political, and social bigotry. Max-Pol Fouchet who writes the forward states:

"It is due to Jean Wahl that such a project became a reality. Without his zeal, this sign wouldn't have happened so early. . . . This project began in the summer of 1942. . . . This homage was undertaken to prove with an act that French thought was on the side of those that defended freedom and the freedom of thought "tout court." . . . Secondly we wanted to show . . . that the United States, too often considered as the land of Ford or Hollywood, possessed an impressive literature, and that the tradition of Washington Irving, Edgar Poe, Walt Whitman, Herman Melville and so many others continues. Also to prove that such an excellent democratic climate for life and spirit, compared with totalitarian countries in which intellectuality was made to be constrained to exile or servitude, permit this blossoming and proliferation of the individual's creative faculties." (1943)

In this same spirit, Wahl emphasizes and affirms the "democratic" dimension of James's radical empiricism in his text "*Les Philosophes Pluralistes.*" This "democratic" philosophy insists on "tolerance" and not only values differences, exchange and flux but requires that one enters the "chaotic" mix and movement of diversity in order to strengthen and enrich one's "moral character." If one partakes in a process of psychic *compenetration* with all sorts of variations of thought, energies, and particles, in "*the most intense game of contrasts and the largest diversity of characters,*" one grows "ethically" stronger, broader, richer and more comprehensive. Only by delving into this "democratic" process does the individual become one of the *free creators* that potentially contributes to "save the world" or generate better worlds.[6]

"A radical pragmatist, says James, is a kind of anarchist, a being who lives without rules, "A la va comme je te pousse."(Moving forward as I push you.)" Since essences and things are disseminated in time and space, it is in their dissemination and their vicissitudes that the person

> eager for this very plenitude of life grasps them. . . . Therefore one has his love for things in their becoming and their flux, his vision of a universe "sub-specie temporis," in its incessant newness. There are infinite reserves of power inside ourselves, and James insists on these innumerable possibilities. . . . The final goal of our creation, James says, seems to consist in the greatest enrichment of our ethical conscience through a game of contrasts in the most intense and the largest diversity of characters. The pluralist will therefore be necessarily tolerant . . . naturally a "democrat";. . . . Democracy must not result in the destruction of the individual's initiative; in fact the world is saved in individuals and by them . . . Such is the religion of democracy of which James speaks . . . above all taken by individuality and that admirable energy of free creators. One can even find here and there in James that which can be an indication of an exterior political program."

> « Un pragmatiste radical, dit James, est une espèce d'anarchiste, un être qui vit sans règle, « à la va comme je te pousse. » Puisque les essences des choses sont disséminées dans le temps et dans l'espace, c'est dans leur dissémination et leurs vicissitudes qu'il faudra que l'homme avide de la plénitude même de la vie les saisisse. . . . De là son amour des choses dans leur devenir et leur flux, de la sa vision de l'univers *sub-specie temporis,* dans sa nouveauté incessante. Il y a à l'intérieur de nous des réserves infinies de puissance, et James insiste sur ces possibilités innombrables. . . . Le but final de notre création, dit James, parait bien consister dans le plus grand enrichissement de notre conscience éthique à travers le jeu de contrastes le plus intense et la plus grande diversité des caractères. Le pluraliste sera donc naturellement tolérant . . . naturellement démocrate. . . . La démocratie ne doit pas aboutir non plus à la destruction de l'initiative individuelle; de même que le monde est sauvé dans les individus et par eux. . . . Telle est la religion de la démocratie dont parle William James . . . avant tout éprise d'individualité et d'énergie, admiratrice des créateurs libres. On pourrait même trouver ça et la dans les écrits de James ce qui pourrait être l'indication d'un programme de politique extérieure. » (Philosophes Pluralistes, 155–156)

Jean Wahl's interest in art and philosophy that potentially "democratize" must be considered in light of his particular position as a Jewish intellectual living and writing in a strong, anti-Semitic climate that finally culminated in Hitler's fascist machine. Jean Wahl obviously had a particular sensitivity to what it meant to create as a minority in a hostile environment and surely had an acute appreciation for the "democratic" principles inherent in James's philosophical expressions.

Wahl's democratic emphasis influences the perspective of this reading/writing that moves between Deleuze's thought and American "rhizomatic" literary texts. In the context of this work it extends the "*democratic*" sense of Wahl's interpretation of James's philosophy and "thinks" it in terms of a "*devenir-democracy*" that may be described as an idealized aesthetic-philosophical *micro-political* process. This use of the term "democracy" should not be mistaken as a metaphor for any actual, existing political regime, a "democracy" as such. This "devenir-democracy" connotes an irreducible dynamic, wherein creative, activating, heterogeneous, conjunctive parts, particles, in flux, and flow generate an un-systematic system that may be thought of as a "democracy in the making." This "concept" helps translate this text's application of, and experimentation with Gilles Deleuze's "pragmatic" or "rhizomatic" enterprise and helps clarify what this reading views as its own difference from interpretive acts it considers "totalitarian" or "anti-democratic."

In the devenir democracy nothing is fixed, or dependent on a "whole" structure that defines positions, relegates, and governs. The parts, particles, molecules move in perpetual flux with infinite differences to constantly generate new differences. According to Deleuze, the writer never creates a world already there waiting to be created, therefore the world it gives life to should never be read in terms of pre-existing ideological concepts of the *real* world, and its adjacent organizational structures, truths, hierarchies, and binary oppositions. A democracy in this sense operates as a process that perpetuates the movement, the creative exchange and evolution that grows out of inter-mixings among different particles or parts that comprise it so that the democracy itself is constantly being renewed, is never a whole, static structure.

Moreover the *devenir-democracy* refrains from categorizing its active participants because these creative individuals engage in the flow and flux of differences to constantly enter a variety of becomings. Therefore to categorize and read them in terms of any stable, unified core identity, necessarily results in gross reductions and distortions. "There are no more forms which are organized in terms of a structure or which develop in function of a genesis; there are no longer subjects, people or characters that let themselves be assigned, formed, or developed. There are only particles left." (*Il n'y a plus de formes qui s'organisent en fonction d'une structure, ni qui se développent en fonction d'une genèse; il n'y a pas davantage de sujets, personnes ou caractères qui se laissent assigner, former, développer. Il n'y a plus que des particules.*) (Dialogues, 146). According to Wahl's reading of James's "democratic philosophy," it destabilizes all notions of given hierarchies, each element, particle, molecule, entity holds an equal value. In the flux and intermingling of these parts each moment mediates new visions, ingenuous contemplations, original thoughts, and new worlds, "worlds in the making."

> "For James any theory of an exterior world that annihilates the idea of exteriority will be necessarily inadequate. . . . And in this world, there is no longer an inflexible hierarchy; each being holds equal importance; each thing carries the same rank. . . . The philosophy of the particular is a democratic philosophy. . . . Thanks to this vision of things, each moment appears to us as a new universe; there is something original, young in nature; we will have returned to the ingenuous contemplation of things, to the first appearance of the world." (PP, 106, 107)

> « Pour James, toute théorie du monde extérieur qui annihile l'idée d'extériorité, sera nécessairement inadéquate. . . . Et dans ce monde, il n'y a pas de hiérarchie immuable; chaque être à une égale importance; chaque chose est sur le même rang. . . . La philosophie du particulier est une philosophie démocratique. . . . Grâce à cette vision des choses, chaque moment nous apparaîtra comme un univers nouveau; il y aura quelque chose d'original, de jeune dans la nature; nous serons revenus à la contemplation ingénue des choses, à la première apparence du monde. » (Philosophes Pluralistes, 106, 107).

Again, the conceptualization of a "devenir-democracy" arises from this textual space "between" William James, Jean Wahl, Gilles Deleuze and the American literary texts concerned. Deleuze does not often use the term "democracy" except for an interesting passage in his essay, "Bartleby, or the Formula" when he speaks precisely of American literature's "democratic" function, that is intricately dependent on its "pragmatism."[7] The use of "Devenir-democracy" in this introduction hopefully helps characterize the perspective of this reader that relates to the potential results of readings and writings that approach thought, and aesthetics in a "pragmatic" fashion. This allusion to a *democratic* demarche where the parts generate malleable, never completed or fixed, in a sense non-whole "whole," also serves to distinguish this textual approach as distinct from those that interpret from the outside in or from the top down. This term is useful because it forces us to question two anti-democratic models often associated with democratic governments. This reading as *democratic* in the sense of the pluralistic process of James hopes to push against these models, as they curtail the potential for a *devenir-democracy.* What are these models? The first is that of the "melting pot," the second is that of "multiculturalism," or "identity politics." The first paradigm advocates a *homogeneicising* of its population, to level or ignore different, cultural heritages until everyone "speaks the same language." The latter scenario *essentializes,* and labels individuals with identification tags that very often confuse physical attributes or traits with cultural background. The latter scenario often means well and does help generate a sense of pride in and acknowledgement of the various contributions made by various ethnic or national groups. The problem comes however with the assumption that those who "genetically" belong to these groups inherently hold a predisposition and knowledge of the culture associated with the group. Similarly it is easy to make the reverse false assumption that those who "genetically" don't appear to belong to the group have no inside knowledge or investment in cultures of other ethnic groups. In other words these groupings make the mistake to base cultural difference with the most superficial aspects of a person. One assumes that individuals from certain geographical locations, or those who have the same skin color, eye shape, religion, nationality, gender, share in one unified "cultural" heritage and identity.[8]

This sense of "devenir-democracy" valorizes elements of diversity, difference and pluralism but underlines that an individual artist's aesthetics should not be read and reduced in terms of any idealized, unified, stable identity which alone gives life to the work. All creative artists breathe life into their works via routes in and out of the flux, flow and chaotic exchange of a plethora of unidentifiable, diverse entities. The "differences" that result from such enterprises are lost in interpretive methods that read works into pre-fabricated ideas that stem from ready made associations between bloodlines, race, body or face, "major history" and intended meanings.[9]

Again, the "devenir-democracy" does not refer to an actual, existing political structure or regime, but does stem from ideas advanced by James and Deleuze that concern the potential power of "micro-political" movements. A *micro-politics* or "devenir-democracy" also informs a particular way of reading, of listening, hearing, and discerning. It breaks with interpretive approaches whose "major," "actual" and "present," "political" emphases, process and proceed in a manner that only reinforces and affirms the categories, striating grids, *binary machines,* and determinate paths that American "minor" or "rhizomatic" literary works undermine, destabilize and potentially dismantle. In other words, through *Deleuze*'s "concepts," semantic signifying fields, traces, particles and agencements, one enters and extends a "hyper-space" that allows for a discussion of texts, which involves issues of race, gender and minority experiences yet avoids becoming and generating other "micro-fascist" forms and formulas.[10] This text contends that current trends of "identity politics," whether with intent or not close the "micro-politics" of aesthetic products down as they read literary *lines of flight* into pre-established, "real-world" political agendas. As Deleuze writes,

> "For what can be done to prevent the theme of a race from turning into a racism, a dominant and all-encompassing fascism, or more simply in an aristocratism, or into a sect and a folklore, micro-fascisms?. . . . The race-tribe exists only at the level of an oppressed race, and in the name of the oppression it suffers: there is no race but inferior, minoritarian; there is no dominant race; a race is not defined by its purity but rather by the impurity conferred upon it by a system of domination. . . . Rimbaud said it all on this point: only those can authorize themselves as of a race who say: "I always was from an inferior race, (. . .) I am from an inferior race, (. . .) I am from an inferior race for all eternity, (. . .) here I am on the "Amoricaine" shores, (. . .) I am a beast, a Negro, (. . .) I am from a far away race, my forefathers were Scandinavians." (Mille Plateaux, 470)

Ross Posnock's book, *Color and Culture,* identifies traces of racist, Jim Crow separatist, ideologies operating at the heart of identity politics. He characterizes this gesture as a genealogical descendent of Booker T. Washington's political platform that "tacitly posited" or reaffirmed the pre-existing "racial ideology of authenticity."

> "And the Bookerite forces, it is fair to say, won the battle over authenticity," that is, the right to legislate who and what counts as truly black and white. . . . Washington stigmatized *black intellectual,* making it a locus of racial, sexual, and economic anxieties. In the late 1960s, Washington's contempt of the intellectual as inauthentic was revived by Black Power and Black Arts nationalism. In our own day, the ideology of authenticity is enshrined as identity politics, the dominant form of multiculturalism." (Color, 16)

Posnock links the legacy of identity politics and multiculturalism to black activists that historically relied on a discourse of inherent, racial difference as a counter productive tactic employed as a means to achieve social equality, and valorization based on race distinctions. Today one wonders why many intellectuals on both sides of the "color line," continue to practice and preach identity politics. Despite multiculturalism's professed intention to value and applaud each group's distinctions and historical contributions to

American culture and history, this realm of discourse despite itself reaffirms, reinforces what Deleuze calls "Major lines" that separate, fix, cut, classify and coerce individuals into pre-designated paths. One doesn't have to look far for examples of this. One need only enter a book store and see its multicultural dividers, *Black* literature, *Black* history, *Native American* history, and the like; We remain caught in a separatist game that also reflects an insidious, subliminal hierarchical structure handed down from a common "white history" that reaffirms the groupings, the *WEs* and the *THEYs,* the *ours* and the *theirs:* As in George Bush the Second's republican convention speech "*We* are *their* America too." Our *Major* history, their *minor* as in less than history: One need only consider those few shelves if one happens to be a curious sort, oddly concerned with the culture and history of "minority life." Otherwise the "Major history" that leaves out "minor" accounts remains the standard. Another example of this may be found in certain forums or documentaries that call in experts to explicate or comment on a minority artist's life and work: The PBS documentary that concerned Ralph Ellison (Spring 2002), only included black scholars such as Lawrence Jackson, Cornel West, and Amiri Baraka (LeRoi Jones). Does Baraka, who calls Ellison a "snob" because Ellison refused certain black nationalist paradigms and characterized his artistry as the product of a miscegenated cultural heritage, really better understand the intention and implications of Ellison's art "*of betrayal*" because they were both born "black" and knew each other? Did Friedrich Nietzsche's sister raised in the same family, another German "intellectual," automatically understand and respect her brother's philosophical intentions? Nietzsche's words perfectly define this misconception. Understanding or not has nothing to do with origins, genes, a shared history or race, but with "psychic" experiences and related predispositions. "Ultimately, nobody can get more out of things, including books, than he already knows. For what one lacks access to from experience one will have no ear. . . . Whoever had understood nothing of me, denied that I need be considered at all."(Ecce Homo, 261).

Deleuze's description of the lines that place people according to binary machines resemble many of our own ideological, intellectual, critical maneuvers that extend, reify and perpetuate these segments that "cut" us up in all sorts of ways, without the intervention of much scrutiny.

> "The segments depend on binary machines, diverse according to need. Binary machines of social classes, sexes, man-female, ages, child-adult, races, white-black, of private and public sectors, of *subjectivations,* from here-not from here."

> « Les segments dépendent de machines binaires, très diverses au besoin. Machines binaires de classes sociales, de sexes, homme-femme, d'âges, enfant-adulte, de races, blanc-noir, de secteurs, public-privé, de subjectivations, chez nous-pas de chez nous. » (Dialogues, 156)

Contrary to an identity politics schema, a Deleuzean field of bodies without organs or faces, forwards that there is no "unified" "subject" as we think we "know" it. A subject never retains an "individualized," unified, identifiable nature. In a Deleuzian field of meaning the "proper name" assumed to represent "one," is always already a "collective" comprised of an innumerable quantity of "agencements" or "relations." Furthermore as these agencements circulate via "contagions," "epidemics" "the wind," rather than being born of genetic codes or descendance, they perpetuate a flow and flux of becomings, "les devenirs," which deterritorialize and disrupt categories imposed by binary machines. As Deleuze writes, the molar order and its corresponding binary codes can't *bite* the molecular currents and lines these agencements produce because they establish liaisons, between heterogeneous elements that include different ages, sexes, reigns, and natures. These "little lines" no longer respond to "molar oppositions" not because they combine to become one, but because they always introduce a "third that comes from elsewhere and upsets the *binarity* of the two." (Dialogues, 158)

In the second and third chapters, this writing indicates certain textual and stylistic literary techniques and improvisations that generate these mutant molecular *lines of flight* that move between the hard molar lines that striate as it considers the texts: *The Souls of Black Folk* (W.E.B. Du Bois), *Melanctha,* (Gertrude Stein), *Cane* (Jean Toomer), and *The Sound and the Fury (*William Faulkner). The movement of these lines creates an *in-between space* that generates the flux of deterritorializations and devenirs-minoritaires. This reading hopes to hear, see, receive and relay some of the immaterial signs, a-signifying sounds, and *differences* at work in the texts that transform metaphors into active metamorphoses. These elements of flux disrupt and disturb the "totalizing, totalitarian," *One* World organizational plan where lines remain subordinate to the points and the points to the *arborescent* system.

> "One can fix a first state of the line . . . the line is subordinate to the point . . . the space that it traces is that of *striage;* the countable multiplicity that it constitutes remains submissive to the One. . . . These kinds of lines are molar, and form an *arborescent* system, binary, circular, segmented. The second kind is very different, molecular and of the "rhizome." The diagonal liberates itself, breaks or winds. The line no longer

makes a contour and it passes between things, between points . . . the multiplicity that they constitute is no longer subordinate to the One . . . the multiplicities of devenir, or of transformations. These are not of numerable elements and ordered relations; but of fluid non-exact elements. From the point of view of pathos, it is . . . schizophrenia that expresses these multiplicities. From the pragmatic point of view, it is sorcery that conducts and handles them."

« On peut fixer un premier état de la ligne . . . la ligne est subordonnée au point . . . l'espace qu'elle trace est de striage ; la multiplicité dénombrable qu'elle constitue reste soumise à l'Un. . . . Les lignes de ce type sont molaires, et forment un système arborescent, binaire, circulaire, segmentaire. La seconde espèce est très différente, moléculaire et du type « rhizome. » La diagonale se libère, se brise ou serpente. La ligne ne fait plus contour, et passe entre les choses, entre les points . . . la multiplicité qu'elle constitue n'est-elle plus subordonnée à l'Un . . . de multiplicités de devenir, ou à transformations, et non plus à éléments dénombrables et relations ordonnées; des ensembles flous, et non plus exacts, etc. Du point de vue du pathos, c'est . . . la schizophrénie qui exprime ces multiplicités. Du point de vue de la pragmatique, c'est la sorcellerie qui les manie. » (Mille Plateaux, 631)

The second chapter reads W.E.B. Du Bois's *Soul's of Black Folk* to specifically respond to an interpretation that "recognizes," relegates and reduces its elements to fit into the "arborescent" Molar system's, politics, racially polarizing structures, and history, to consequently nullify the "micropolitical" lines and becomings of the text. This contested interpretation provides an example of what Deleuze refers to as a "*reterritorializing*" reading in that it forces the mutant lines that move between and beyond the Molar System's organizational schemas and binary machines back into a context or terrain of "logical" sense. Again, such interpretations either tie the lines back into the points that move only subordinate to the *One,* or simply efface, ignore or miss the a-signifying non-material signs or sounds. Deleuze emphasizes that this *One* "belted in," *englobing,* codifying conceptual world snuffs out the devenirs, *les lignes de fuite* in its organizational, "mortuary" reterritorializing process. « La terre ceinturée, englobée, surcodée, conjuguée comme objet d'une organisation mortuaire et suicidaire qui l'entoure de partout. » *The belted in Earth, enclosed and encoded, conjugated as an object of a mortuary and suicidal organization which surrounds it on all sides.* (Mille Plateaux, 636)

To appreciate without reducing mutant and mutating lines of flight requires a particular textual approach. Deleuze juxtaposes two types of *plans.* The first one named "organization," orders and insists on the forms and formations of subjects in terms of structure and "genetics." This "plan" designates each "agencement" with a single meaning to result in reterritorialization. The second one, *a plan of consistence* or *immanence* does not "know" what the former plan conceptualizes as "subject" but instead values the "agencements" or "hecceites": This plan's "pragmatic" or "diagramic" component liberates the multiplicities inherent in the agencements: their pieces and lines that have been encoded and plugged to perpetuate further mutations and conjugations. (Dialogues, 141)

An "organizational plan" that coerces the text into an exterior terrain of "real politics," "real history," "real intellectual logic," and its author into a "real" genetically, historically traceable, unified identity, only destroys the micro-political, molecular particles and forces that the writing potentially proliferates. Contrary to reterritorializing interpretations, the "sorcellerie" of the "pragmatic" approach activates and conducts, abnormal nomadic multiplicities, *multiplicités de devenir, ou des transformations, des ensembles flous, et non plus exacts (multiplicities of becoming, or transformations, the fluid ensembles, no longer exact),* to again create rather than critique. There are no more "unified" raced, gendered, historicized subjects or regulatory laws and judgments to be made; there are simply agencements to be entered into and new lines to be extended. The *devenir-democracy* only reaches its potential for "micro-political" efficiency, if the "reader" becomes writer rather than critic or judge. Deleuze characterizes this activity as an encounter that sweeps more particles into the "wind" to disseminate and generate active, creative mutant lines of flight that actualize more "agencements," more heterogeneous, anti-hereditary, non-personal alloys and alliances. "Find, meet, fly/steal a-way, instead of regulating, recognizing and judging. Because recognizing is the opposite of meeting. . . . It is better to be a sweeper than judge . . . advance a line or a block between two people." « Trouver, rencontrer, voler, au lieu de régler, reconnaître et juger. Car reconnaître, c'est le contraire de la rencontre. . . . Plutôt être balayeur que juge . . . faire filer une ligne ou un bloc entre deux personnes. » (Dialogues, 15, 16)

This reader tries to "meet," "receive," these texts in such a manner so as to "conduct" and extend their multiplicities, particles, lines, but also to make the case that a pragmatic approach not only "unplugs" lines, "stuck" in organizational structures or readings, but also points out the "organizational properties" of certain "major" aesthetic products that also subordinate lines to make *"major"* points and consequently reify such structures. In light of this issue the third chapter considers the distinction Deleuze makes between writings that "betray" (*livre-rhizome/ book-rhizome*) the One unified world order, and writings that "cheat" (*livre-racine/ book-root*) to reinforce "molar" world structures. The latter kind of writing for the most part reflects and mediates preconceived concepts, scenarios and ideologies. "A first type of book is that of the livre-racine (book-root). The tree is already the image of the world, or rather the root is the image of the tree-world. This is the classic book. . . . The law of the book, is that of reflection. . . . The binary logic is the spiritual reality of the tree-root." « Un premier type de livre, c'est le livre-racine. L'arbre est déjà l'image du monde, ou bien la racine est l'image de l'arbre-monde. C'est le livre classique. . . . La loi du livre, c'est celle de la réflexion. . . . La logique binaire est la réalité spirituelle de l'arbre-racine. » (Mille Plateaux, 11) This kind of fiction much in the manner of identity politics bases its assessments, observations and interpretations on already designated, differences and definitions and injects its "reflections" apodictically into the writing. This manner of "writing" that Deleuze characterizes as "cheating" only serves to reinforce the fixed powers of the earth that want to situate us and curtail our movements. On the contrary, the "experimenter's" writing that "betrays" creates a *smooth space* free of grids, or striating lines, while it extends lines of flight, individuations and agencements that interact independent of history, time, "majority" rules, points and segments, to finally undermine and diminish the legitimacy of "fixed properties." Deleuze expresses that there is always "betrayal" in a "line of flight," the betrayal of one who no longer has any past or future. (Dialogues, 52)

The pragmatic "philosopher" reader/writer also has to create as a "traitor," a traitor to theories and ideologies that polarize the fields of art and interpretation. One must break with interpretive enterprises that inflict pre-existing schemas, "rational," truths and concepts on "fantasy space" to make it "meaningful." To provide a concrete example of what is meant here one need only refer to a recent international conference on the "usefulness" of pragmatic approaches in literary studies. A synopsis of the conference summarizes one participant's paper of opposition "*Why Pragmatism is not Very Useful for the Arts*" in the following terms: "Altieri held that because the strength of pragmatism lies in providing alternatives to the insistence on strict criteria for "truth" it cannot offer much in the way of useful concepts when the area involved is already resistant to languages of truth and in a context where theory has to be able to deal with fantasy space." According to this summary Charles Altieri fails to see the value of a pragmatic approach because he views its operations and goals in terms of those of standard critical interpretive procedures. However pragmatic thought doesn't think in a manner which positions "truth languages" here and "fantasy languages or space" over there. *Pragmatic* "truths" move through language in an inextricable process of flux and flow that resist methods that reduce them to a terrain of reality, or logical space. It does not follow that because pragmatism resists "truth languages" that stem from ontological foundations, that it necessarily fails to consider and reveal 'truths' inherent in *fantasy space.* In fact because the "pragmatic" process esteems the most forceful truths as operating, moving, flowing in "fantasy spaces" it instead activates, drives these truths on, but on a mutually constructed deterritorialized terrain. David Lapoujade, a disciple of Deleuze writes in his book on Jamesean pragmatism:

> "In fact pragmatism proposes less a new definition of truth than a method of experimentation, of construction of new truths. To experiment is to consider theory as a creative practice. This is why it is not about knowing if it is "true," but knowing how the truth makes itself. . . . The true idea, this is not only what one believes, what one does, or thinks; this is what makes one believe, that which makes one act or that which makes one think. . . . Therefore if truth is action, transition, creation (rather than representation, conclusion, imitation), this is to the extent to which "the complete truth is the truth that gives energy and delivers battles."

> « A la lettre, le pragmatisme propose donc moins une nouvelle définition de la vérité qu'une méthode d'expérimentation, de construction pour de nouvelles vérités. Expérimenter, c'est considérer la théorie comme une pratique créatrice. C'est pourquoi il ne s'agit plus de savoir ce qui est vrai, mais comment se fait le vrai. . . . L'idée vraie, ce n'est pas seulement ce qu'on croit, ce qu'on fait ou ce qu'on pense; c'est ce qui fait croire, ce qui fait agir ou ce qui fait penser. . . . Si donc la vérité est action transition, création (plutôt que représentation, conclusion, imitation), c'est dans la mesure ou « la vérité complète [est] la vérité qui donne de l'énergie et livre des batailles. » (Wjames, 59, 60)

It would seem the implications of Altieri's words would be that the only way to "render" the "truth" of fantasy space, "real" or "understandable," is to translate it into the rational language of ontological "truths" and concepts. But in that the discursive truths of "fantasy space" also resist *languages of truth,* to remove them from their aesthetic, poetic context and to translate them into graspable, conceptual terms, only aborts the original life forces, lines, intensities that once were there. The value of the pragmatic approach forwarded by Deleuze lies in the fact that irreducible truths, "immaterial" essences get extended on due to the fact that one reads/writes, creates, "believes," "activates" *with* them and through them, rather than "rationally" writes "about" them. Perhaps nobody more than literary artists such as Baudelaire, Nietzsche, Maupassant, Virginia Woolf, who find their "fantasy-space-truths" violated time and time again in *informative* "interpretations" better express the "reality" of such rational, reductive renderings.[11]

> "One must discuss and contest their theory that one can summarize in the following words, "nothing but the truth and only the truth." . . . What childishness, to believe in "reality" as each one of us carries our own in our thought and our organs. Our eyes, ears, our senses, our different tastes create as many truths as there are men on the earth. And our spirits which receive instructions from these organs, diversely affected, understand, analyze, and judge as if each one of us belonged to a different race. . . . The partisans of objectivity (what a villainous word!) pretend on the contrary, to give us an exact representation of that which takes place in *life.*"

> « On doit discuter et contester leur théorie qui semble pouvoir être résumée par ces mots, « Rien que la vérité et toute la vérité » . . . . Quel enfantillage, d'ailleurs, de croire à la réalité puisque nous portons chacun la notre dans notre pensée et dans nos organes. Nos yeux, nos oreilles, notre odorat, notre goût différent créent autant de vérités qu'il y a d'hommes sur la terre. Et nos esprits que reçoivent les instructions de ces organes, diversement impressionnées, comprennent, analysent et jugent comme si chacun de nous appartenait à une autre race. . . . Les partisans de l'objectivité (quel vilain mot!) prétendent, au contraire, nous donner la représentation exacte de ce qui a lieu dans la vie. » (Pierre et Jean, 4, 5).

To unplug, un-code, liberate the lines, pieces, particles that get "organized" in "developmental plans" one has to "sympathize with," think of the author so strongly that they cease to be either "subject" or "object." "L'agencement, is the co-functioning, it is the "sympathy," the symbioses. . . . Here other devenirs interconnect . . . seized in their particles at the same time as their flux to conjugate with my own: An entire world of micro-perceptions which lead us to the imperceptible."(Dialogues, 60, 65). Deleuze writes that this experience has everything to do with experimentation and nothing to do with interpretations. Only if the reader enters the deterritorialized terrain of the "livre-rhizome," or the writer's "regime of signs" and conjugates with and appreciates its micro-political movements, elements, agencements can he/she return the joy, force, "vie amoureuse" and *politics* that the writer knew how to invent and give. Deleuze tells us his writings *with* Kafka or Proust, move in this direction so as not to *betray* the writer's life, not to cause any sadness, "anything that would make the writer cry in the grave." (Dialogues, 141, 142)

As this text considers "Deleuze's "regime of signs" and all the other names included in this writing it also "hopes" to extend their *lives, loves, politics,* that they knew how to produce and offer. As already mentioned this reading/writing should be received as a "double gesture" that generates one interactive, *involute* pragmatic space.[12] While this textual tracing maps out a certain *plan of consistence,* where the lines of both literary artists and pluralist philosophers that *make/do* a "pragmatics," run, intersect, and move on, it hopefully activates and *futurizes* their multiplicities and mutant, micro-political lines that Jean Wahl tells us potentially "save worlds," or create better worlds. The conviction in the efficacy of such an enterprise demands an irrational *faith* and a certain obstinacy. The question that begins with "EST" (IS) must be replaced with the process or "art of the 'ET' (AND)." Instead of asking "*Is* this really *micro*-politically effective?" "*Is* this simply romantic idealism, folly or absurdity?"; one must advance sense-less, blind, deaf and dumb to the organizational plans and "rational" notions and ideals of the "real" *One* world and its adjacent interpretive methods. Instead one must move actively to co-construct the milieu of the *radicals, with* "les clefs dans le vent," in/*with* the spirit of William James, Jean Wahl and Gilles Deleuze.

> "A radical pragmatist says James, is a kind of anarchist, a being who lives without rules, "In the movement as I push you.." . . . And finally this philosophy is therefore willfulness "From the breast of man rises a free and strong determination in his language and his mind. And he cries to himself in spite of his nature and his milieu: "I want!." . . . In this pluralist philosophy . . . the world is an ensemble of lives which can reach the greatest degree of complexity possible, infrahuman, *suprahuman,* as well as human, evolving and changing profoundly in their efforts and attempts and in their interactions and their accumulated successes, all this composes the universe."

> « Un pragmatiste radical dit James, est une espèce d'anarchiste, un être qui vit sans règles, « *a la va comme je te pousse.* » . . . Et enfin cette philosophie est donc un volontarisme « De la poitrine de l'homme à sa langue et à son cerveau monte une libre et forte détermination. Et il crie de lui-même, et en dépit de toute sa nature et de tout son milieu: je

> veux!. . . . Dans la philosophe pluraliste . . . le monde est pour elle un ensemble de vies qui peuvent être de tous les degrés de complexités possibles, infrahumaines, suprahumaines, aussi bien qu'humaine, évoluant et changeant profondément dans leurs efforts et leurs essais, et par leurs interactions et leurs succès accumulés composent l'univers. » (Philosophes Pluralistes, 112, 114, 161)

Furthermore as Deleuze warns us, each additional moment one stays fixed in the question *Is* the revolution possible?" That many more people miss becoming revolutionary. "This question is made precisely for that reason, to prevent a devenir-revolutionary of people, at every level, at every site." One must replace the pervasive cynicism that infects us with the notion that "the revolution is impossible," with the active thought that thinks a new kind of revolution into the process of *becoming* possible, a *becoming*-revolutionary.

> "Instead of betting on the eternal impossibility of the revolution . . . why not think that a new type of revolution is in the process of becoming possible, and that all sorts of lively mutant machines, lead wars, conjugate, and trace out a plan of consistence which undermines or guts out the organization of the World." (Dialogues, 173, 176)

Chapter One

France's "Two Most Important Philosophers": Jean Wahl and Jean-Paul Sartre

> "A part from Sartre, who remained caught nonetheless in the trappings of the verb to be, the most important philosopher in France was Jean Wahl. He not only introduced us to an encounter with English and American thought, but had the ability to make us think, in French, things which were very new: he on his own account took this art of the AND . . . this minoritarian use of language, the furthest." (Dialogues II, 57).

Deleuze's adulatory terms applied to American literature, continue on or inspire the works and words of post modern French authors such as Dantec. However, it is a naïve assumption to believe that Deleuze is the starting point of these positions. Deleuze himself always emphasizes that there is never any one true localizable starting point to any one thought or for that matter group of thoughts. Gilles Deleuze like all thinkers, finds inspiration in a variety of works, words and movements of innumerable writers and artists who precede him to eventually create and produce in that way which is so uniquely and authentically "Deleuze."

In keeping with one of Deleuze's greatest influences being Friedrich Nietzsche, it is only out of "inspiration," which has nothing to do with mimesis but everything to do with creating something *different,* that innovative, life bearing creations arise.

> "Has anyone at the end of the nineteenth century a clear idea of what poets of strong ages have called *inspiration?.* . . .—If one had the slightest residue of superstition left in one's system, one could hardly reject

altogether the idea that one is merely incarnation, merely mouthpiece, merely a medium of overpowering forces. . . . One hears, one does not seek; one accepts, one does not ask who gives; like lightening, a thought flashes up. . . .—I never had any choice . . . an instinct for rhythmic relationships that arches over wide spaces of forms—length, the need for a rhythm with wide arches, is almost the measure of the force of inspiration, a kind of compensation for its pressure and tension. . . . ("Here all things come caressingly to your discourse and flatter you; for they want to ride on your back. On every metaphor you ride to every truth. . . . Here the words and word-shrines of all being open up before you; here all being wishes to become word, all becoming wishes to learn from you how to speak." *Zarathustra III*)." (Ecce Homo, 300, 301).

Deleuze never forgets Nietzsche's recounting of his experience of inspiration or fails to use it and its adjacent doctrine of the eternal return to push for a dismantling of philosophical approaches that refuse to break out of Cartesian and Platonic metaphysical foundations. Deleuze employs Nietzsche's "little ditty," the refrain of the eternal return to explain the process of cosmos philosophy that transmits mute and unthinkable forces and mobilizes them ever forward in a rhythmic system that allows for non-material molecular particles to fuse in heterogeneous space times and generate new, becoming differences or "milieus" that perpetuate this creative, generative enterprise. "Let us recall Nietzsche's idea of the eternal return as a little ditty, a refrain, but which captures the mute and unthinkable forces of the Cosmos."(*Thousand Plateaus,* 343).

"The notion of the milieu is not unitary: not only does the living thing continually pass from one milieu to another, but the milieus pass into one another; they are essentially communicating. . . . What chaos and rhythm have in common is the in-between—between two milieus, rhythm-chaos or the *chaosmos.* . . . There is rhythm whenever there is . . . a communication of milieus, coordination between heterogeneous space-times. . . . This is Cosmos philosophy, after the manner of Nietzsche. . . . The forces to be captured are no longer those of the earth . . . but the forces of an immaterial, nonformal, and energetic Cosmos. . . . Philosophy is no longer synthetic judgment; it is like a thought synthesizer functioning to make thought travel, make it mobile, make it a force of the Cosmos (in the same way as one makes sound travel)." (Thousand Plateaus, 313, 314, 342, 343).

Deleuze's interpretation of Nietzsche's eternal return characterizes a philosophical operation that cannot be described in ontological terms. This cosmos-philosophy *captures the mute and the unthinkable forces of the Cosmos:* the immaterial, nonformal, energetic forces and disperses them in its rhythmic movements. These molecular deterritorialized materials exceed the concepts of *romantic* philosophy that thinks in terms of chronological progression and identifies only what it perceives as part of its system of absolute essences and truths. Conversely, "Romantic thinking" forces the rhythms, into a schema that only wants to reinforce its concepts and theories and consequently renders all that moves beyond its ideas, those chaosmos creations of the non-material sort, into distortions and deformations of what they once were. Deleuze further speaks of the rhythms of "*chaosmos philosophy*" as "agencements" in *Dialogues.* In this text, Deleuze describes "agencements" as multiplicities which include a limitless number of heterogeneous terms. These terms found liaisons and relations between them, relations of various ages, sexes, reigns, and natures. The "agencement" does not establish a fixed "unity" but is a mobile and malleable co-functioning: that becomes through "a symbiosis," "sympathy," or a natural attraction that instinctively aligns two entities. The alliances and alloys that join these diverse elements result from encounters fueled by *contagion, epidemics, the wind,* rather than from heredity, or descendance. (Dialogues, 84, 143)

This "chaosmos philosophy" operates unsystematically and spontaneously. It traces out its "deterritorialized" territory as it delves into a whirlpool of countless, energy fields and moves in and out of different milieus or agencements that each carry a multiplicity of heterogeneous, immaterial, non-personal, and non-formal terms that activate intersections with other agencements. These dynamic individuations cannot be logically "known" in terms of ontological paradigms, assignable names or historical periods. Deleuze expressly characterizes his own philosophy as that of the Chaosmos version. In that an artist's or an artist-thinker's creations result from a collective communication with a multiplicity of "particles" that meet in "heterogeneous space times," to interpret a thinker's intellectual progress in terms of logic based, conceptual systems or textbook intellectual history coerces the non-material particles into a materializing system that aborts the innumerable, imperceptible particles, souls, that comprise a "collective" regime of signs. Because Deleuze's *demarche* creates its playing field out of this non-systematic system, this study which primarily applies Deleuze's theories to American modernist texts necessarily includes a plethora of ageless, nameless echoes. One can say with Deleuze that there is no unified subject who is "Deleuze" in his writing; his textual corpus disseminates a combination of forces, entities, energies, names that are themselves fusions, diffusions of a multiple number of other "names" and sign regimes.[1]

This first chapter provides a certain intellectual context for Deleuze's ideas that concern American themes as it considers the influences of two figures that wrote, taught and were active on the Paris scene when Deleuze was a young, philosophy student at the Sorbonne: Deleuze himself names these thinkers "France's most important philosophers." These two men play pinnacle roles in stimulating Deleuze's interest in American thought and American literature. However, keeping in mind Deleuze's interpretation of the eternal return this text warns against misreading or equating this with an attempt to define Deleuze's eclectic project in terms of what these other thinkers write and think. This approach rather considers these sources because they lend a certain vantage point or provide insights into the directions Deleuze takes up and propels forward.

The first of these French philosophers this work reads in relation to Deleuze *being* Jean-Paul Sartre is often assumed to be a counter figure to Deleuze. However, to refer to Sartre does not suggest that Deleuze writes in a Sartrean way or thinks as an existentialist or the like, but instead affirms that Deleuze discovers certain tools in Sartre's "box," ones which amongst others allow Deleuze to write "out of the box" or create a non-boxed set of "tools." One can think of these tools in a pro-genitive sense. One that evokes Deleuze's use of the term rhizome, the most reproductively efficient of all bio-botanical life forms, and his description of the active forces found in an engagement leading to a marriage that produces something "new" and unique. Deleuze's conceptualization of this idea is intimately linked to Nietzsche's doctrine of the eternal return which Deleuze articulates in his publication, *Nietzsche et la philosophie.*

> "In effect, the unequal, the different is the veritable reason of the eternal return. This is because nothing is equal, nor the same, that "it" returns. In other words, the eternal return refers only to the "devenir" (becoming), to the multiple. It is the law of a world without being, without unity, without identity. . . . Nietzsche's secret is that the eternal return is selective. . . . It causes the active forces to affirm, and affirm their difference. . . . Only affirmation, that which can be affirmed, joy, returns. All that can be denied, all that is negation, is expulsed by the very movement of the eternal return. . . . In order that the affirmation be affirmed Dionysos has a fiancée, Ariane. The only wise word is 'yes.'" (24, 37–39).

Deleuze's interpretation of Nietzsche's eternal return insists that after *the marriage* of two entities, the offspring or offshoot becomes as an entirely unique organic life. The important vocabulary in this semantic field includes the signifiers, the unequal, the differand/different, the multiple, not equal, without unity, without identity, nor of one nor the other but of many others, still becoming infinite others is the "law" of writing in the name of this new conceptual view of the eternal return. To extend Deleuze's constant use of reproductive metaphors; the new life carries forward certain inherited genes. However, the "final" result, make-up, or points of origination of the genetic mix remain mysteries that have no logical, verifiable answers. There is no way to decode what percentage of whose genes make up that life and it is always the case that genes from infinite sources, too far back for tracing or *knowing,* are always already in the mix.

In response to Deleuze's description of his early academic writing as a philosophy student one will ask in which manner does Deleuze beget a new life out of Sartre? Deleuze writes in a letter to Michel Cressole that he had to compensate for feeling "bludgeoned to death with the history of philosophy" while a philosophy student at the Sorbonne.[2] He characterizes these early writings as more reactive than those creative acts produced in the non-reactive spirit of the eternal return. The pre-Nietzsche publications are rather violations described in terms as *"taking* an author from behind."

> "I suppose the main way I coped with it at the time was to see the history of philosophy as a sort of buggery [enculage] or (it comes to the same thing) Immaculate Conception. I saw myself as taking an author from behind, and giving him a child that would be his offspring, yet monstrous. It was really important for it to be his child, because the author had to actually say all I had him saying. But the child was monstrous too because it resulted from all sorts of shifting, slipping, dislocations, and hidden emissions that I really enjoyed." (Two Fold, 14–15)

As one considers Sartre's role as thinker in Deleuze's entourage and his writings and life as an intellectual, one moves closer to assessing whether their "engagement" results in a "monster" or a lovelier love-child. What does Deleuze think of Sartre, perhaps the most notable or visible French intellectual figure on the Parisian scene of his formative years as a philosopher/intellectual? The author of *The Two-Fold Thought of Deleuze and Guattari,* Charles J. Stivale, turns to Michel Tournier to provide a sense of Sartre's role in the lives of French students who were academically formed in the wake of Sartre's prolific thought. Tournier, Deleuze's co-student, philosopher and novelist names Jean-Paul Sartre the major figure that influences, and stimulates philosophical action and reaction amongst the younger generation. Tournier relays that Sartre had momentarily freed the younger philosophy students from feeling they had to be formed to be the guardians of what Tournier describes as the philosophical citadels of the past. "In fall 1943 a meteor of a book fell onto our desks: Jean-Paul Sartre's *Being and Nothingness.* After the initial exultation of seeing a philosophy born before our very eyes, our master had gone and fished up that worn-out old duffer Humanism . . . still stinking with sweat and 'inner-life,' from the trash heap where we had left it." (*Two Fold,* 4, 5)

Tournier explains that the humanist dimension of Sartre's philosophy however betrayed the fact that it carried traces of metaphysical tendencies, which caused a subsequent rejection amongst the younger generation. A quote from Althusser also confirms the extent to which Sartre's presence and ideas impacted younger French thinkers. "For much of the work in French philosophy undertaken from the late 1940s onward developed under the shadow and weight of Sartrean doctrine and authority, in response to which the fashion arose among young philosophers to appear to despise Sartre." (*Two Fold,* 5). One of the key words in Althusser's statement is the "appear" in *appear to despise.* Tournier also underlines the superficiality of the *juvenile excess in condemnation.* " . . . a liquidation of the father by overgrown adolescents afflicted with the awareness that they owed him everything." (*Two Fold,* 5)

Stivale seems to enquire into Deleuze's relation to Sartre with everyone in his entourage but strangely enough neglects or overlooks Deleuze's own explicit remarks that communicate the extent to which he revered Sartre. In fact contrary to Sartre being described as an "authority" that causes young philosophers to feel oppressed under the weight of his shadow, Deleuze points to Sartre as a key to his liberation. In fact Sartre plays the role of a sort of "sur place" savior that releases him from the drudgery of the traditional philosophical paths plotted out before him, *a scholasticism worse than that of the Middle Ages,* so that instead of the act of "enculage" he can create organic *becomings* rather than monstrous distortions. Deleuze describes Sartre as the "Outside" and affirms that Sartre never stopped being that, a "gust of air" rather than a model, method or example: "An intellectual who singularly changed the situation of the intellectual." (*Dialogues II,* 12)

Deleuze as do Foucault and Tournier, pays heed to Sartre's influence but contrary to his friends, Deleuze writes that if "existentialism" was something that was already "past history" to be left behind Sartre was not.[3] Deleuze, in turn frees Sartre, who never ceased being that pure air, an *Outside, a draft from the Arrière-cour,* from being summed up by or reduced to the "concept" existentialism. Deleuze's words emphatically imply that although existentialism holds no interest for him; Sartre's philosophical and aesthetic movements do. "Albeit I didn't feel attracted to existentialism at this time, nor towards phenomenology, I really don't know why, but it was already history when we got there, too much method, imitation, commentary, and interpretation, except by Sartre." (Dialogues, 19).

Deleuze characterizes Sartre's significance in very similar terms that Sartre applies to Kafka, the writer par excellence, that writer that renders the irreducible truth of appearances and makes one intuit above them, an *other* truth, which will always be refused us. « On n'imite pas Kafka, on ne le refait pas: il fallait puiser dans ses livres un encouragement précieux et chercher ailleurs. » (One doesn't imitate or re-do Kafka: it is necessary to grasp a precious encouragement from his books than look elsewhere.) (Situations II, 255). Deleuze similarly says of Sartre, that he is not a model, method example to be imitated but rather something to be breathed in as a *precious inspiration* as the *unique combination* that rendered the younger generation strength, *la force.* (Dialogues, 19).

Deleuze then continues on to pay Sartre the highest compliment when he describes him in terms of "rhizome" incarnate when he points out the stupidity in trying to place Sartre as either the beginning or the end of something. "It is stupid to wonder if Sartre is the beginning or the end of something. As all creative things and people, he is of the middle, he springs forth through the middle." « C'est stupide de se demander si Sartre est le début ou la fin de quelque chose. Comme toutes les choses et les gens créateurs, il est au milieu, il pousse par le milieu. » (Dialogues, 17, 18) The introductory chapter to Deleuze's *Mille Plateaux* can help elucidate further what Sartre as "*rhizome*" *incarnate* signifies.

> "A rhizome has no beginning or end; it is always in the middle, between things, interbeing, *intermezzo.* . . . The tree imposes the verb "to be," but the fabric of the rhizome is the conjunction, "and . . . and . . . and." . . . Seeking a beginning or a foundation- all imply a false conception of travel and movement . . . another way of traveling and moving, proceed from the middle, through the middle, coming and going rather than starting and finishing." (Thousand Plateaus, 25)

> « Un rhizome ne commence et n'aboutit pas, il est toujours au milieu, entre les choses, inter-être, *intermezzo* . . . L'arbre impose le verbe « être, » mais le rhizome a pour tissu la conjonction « et . . . et . . . et. » . . . Chercher un

> commencement, ou un fondement, implique une fausse conception de voyage et du mouvement . . . une autre manière de voyager comme se mouvoir, partir au milieu, par le milieu, entrer et sortir, non pas commencer ni finir. » (Mille Plateaux, 36)

What is even more interesting than Deleuze characterizing *Sartre,* as "rhizome" is the fact that this is the term Deleuze consistently applies to describe American literature; that literature that Sartre more than any other French thinker of his generation praises, promotes and tries to emulate. In the book, *Extreme Occident,* Jean-Philippe Mathy documents this. "Sartre was extremely attracted to the American literature of the 1920's and 30's: perhaps more so than any other french writer of his generation. In a delivery at Princeton in 1946, Sartre acknowledges the debt a whole generation of young French writers owed to the writings of Hemingway, Steinbeck, and Dos Passos. "It seemed to us suddenly that we had just learned something and that our literature was about to pull itself out of its old ruts. At once, for thousands of young intellectuals, the American novel took its place, with jazz and the movies, among the best imports from the United States."(Extreme Occident, 130).

One can say in fact that Sartre's writings may be one of the sources that help Deleuze formulate his distinction between a major and minor literature. A few reference points that the two thinkers share in common may help elucidate this. Both thinkers refer to Kafka and American modernist writers as prime examples of artists that create the kind of literature that moves into the future. Deleuze and Sartre also talk about the writer as a political animal first and foremost (albeit in most different ways). These writers also emphasize the importance of aesthetics and in some manner articulate that artistic expressions surpass philosophical, historical and scientific attempts to translate the *real* experiences of human existence. Art speaks to the *whole* being, the mind, spirit, senses and body, whether through sounds, images or silences, and transmits existential *realities* that reason based disciplines efface, reduce or simply can't account for. Although these commonalities in theme exist, that does not mean that the two thinkers would be in accord on all issues and believe in all of the same values or lend the same meaning to the same signifiers.

Let it suffice to say that Sartre writes of these themes first and Deleuze takes them up and writes on them subsequently and *differently.* Sartre's work inspires something unique to evolve out of Deleuze. In other words, although Deleuze's writings echo many of Sartre's themes, Deleuze's conceptual framework shifts the meanings of concepts that make up these themes or replaces those concepts all together. In sum, far from reiterating what Sartre already said Deleuze rather takes up Sartre's thematic road and along with other digested fragments, he blazes his own unique deterritorialized territory.

This text proceeds to ask, what are a few examples of similar themes at work in both Sartre and Deleuzes' works that relate to this study of American modernist works? How do these themes developed differently in each thinker's oeuvre shed light on the uniqueness and particularity of the meanings at work in certain concepts in Deleuze's work? How do Sartre and Deleuze's ideas of the "écrivain engagé" differ? In relation to this, how do they perceive *the political* in the aesthetic product? And, how does contrasting these two thinkers lead one to better understand Deleuze's *differences* and how they operate so that the implications of their meanings will be better understood when applied to American modernist works in subsequent chapters?

Sartre decries the failure of French letters to reach the same innovative status of American literature and blames it on a society that feigns superiority founded on a "dead world" bound down by a historical view that limits and renders its writers defunct. According to Sartre the novels of the prior generations recounted events retrospectively to support "logical perceptions, universal relations, and eternal truths." One already understood the slightest change before it even occurred. Sartre urges the writers of his day to break with these models to instead "provide the event with its brutal freshness, ambiguity and unpredictability." "We do not want to satisfy our public with its superiority over a dead world that we would rather wish to take by the throat."(Situations II, 254).

Deleuze also suggests that the French fail to generate a truly creative, literary form due to the illusions of the "historical" France and its glories which French writers internalize and project into their oeuvres. This overrated image of the Mecca of western civilization, which obsesses on the past and future while it neglects the present, is that of Sartre's "dead world" which feigns superiority. Deleuze like Sartre upholds American literature as an alternative, "superior" model that pushes beyond established historical, cultural, linguistic boundaries, beyond "arborescent" structures, and their corresponding hierarchies and binary machines. Deleuze describes American literature as operating according to geographical lines, to cross frontiers, and push them back in order to move beyond. French writers are described in contrast as stuck in historical concepts and "making points" rather than new lines. "They are too fond of roots, trees, surveys, points of arborescence, and properties." "The becoming is geographical. There is no equivalent in France. The French are too human, too historical, too concerned with the future and the past." (Dialogues, 48)

Many times in Sartre's text *Situations II,* one finds some of Sartre's greatest praise for American letters grouped with his praise for Kafka who Sartre points to continuously as the writer par excellence. Kafka and American modernists create a new literary form due to the fact that they replace old techniques and tendencies that no longer render an accurate expression of the modernist world of flux, change and instability. Sartre calls for a literature that writes without guarantees, and communicates a sense of the chaos and unpredictability inherent in the modern, human condition. French literature must also present life in movement rather than a mummified life that old, outdated methods cast into redundant pre-established scenarios and moral schemas. Sartre ascribes the achievements of American writers to the fact that they create characters that "trap" the reader and throw him/her from one consciousness to another which communicates an "absolute and irremediable universe" that is perpetually renewed, as the present is always side-stepped towards the future. The reader finally "feels each of their moods, that each movement of their spirit encompasses the entire humanity in their time and space." (Situations II, 255). Sartre continues on to hold Kafka a part as beyond imitation, "impossible to redo" while he contrasts Americans and French writers in terms of their historical and geographical conditions. Sartre states that as French writers were overwhelmed and weighed down by their history, Americans were both cursed and blessed in being freed from traditions: Without traditions and historical reference points Americans incurred feelings of stupor and abandonment before incomprehensible events that took place upon a "continent that was too big." However these same unprecedented and difficult experiences demanded that American writers come up with "strange" innovative methods to communicate historically uncharted circumstances and experiences.

> "On can neither imitate, nor redo Kafka. . . . As for the Americans . . . a literature that, in feeling threatened because its techniques and myths would no longer permit it to face up to the historical situation, scrambled for strange methods in order to fulfill its function in new conjectures." (Situations II, 255, 256)

> « On n'imite pas Kafka, on ne le refait pas. . . . Quant aux Américains . . . une littérature qui, se sentant menacée parce que ses techniques et ses mythes n'allaient plus lui permettre de faire face à la situation historique, se greffa des méthodes étrangères pour pouvoir remplir sa fonction dans les conjectures nouvelles. » (Situations II, 255, 256)

Deleuze similarly exemplifies what he means by "livre-rhizome," "la littérature mineure" through literary examples provided by American moderns and Kafka. Deleuze tells us that Kafka and American writers know how to create minor becomings through their literature because they know how to write in minor languages. Kafka writes as a Czech-Jew writing in German, American moderns who dislocate standard forms of English, similarly write as foreigners in their own language.

> "'Major' and 'minor' do not qualify two different languages but rather two usages or functions of language. . . . Kafka, a Czech Jew writing in German, submits German to creative treatment as a minor language . . . stretch the tensors through all of language, even written language, and draw from it cries, shouts, pitches, durations, timbres, accents, intensities. . . . Black Americans do not oppose Black to English, they transform the American English that is their own language into Black English. . . . Use the minor language to *send the major language running.* The minor author is a foreigner in his own tongue." (Thousand Plateaus, 104, 105).

> « « Majeur » et « mineur » ne qualifient pas deux langues, mais deux usages ou fonctions de la langue. . . . Kafka, Juif tchèque écrivant en allemand, c'est à l'allemand qu'il fait subir un traitement créateur de langue mineure . . . tendre des tenseurs dans toute la langue, même écrite, et en tirer des cris, des clamés, des hauteurs, durées, timbres, accents, intensités. . . . Les noirs-Américains n'opposent pas le black à l'english, ils font avec l'américain qui est leur propre langue un black-english. . . . Se servir de la langue mineure pour faire filer la langue majeure. L'auteur mineur est l'étranger dans sa propre langue. » (Mille Plateaux, 131, 132, 133)

French writers disappointed Sartre as they resisted breaking with past models, predecessors, and continued to write from a "historical" perspective. Sartre argues that French writers in trying to escape their *historicity* wrote with a transcendental, superior historical vision, to create little more than *dogmatic realism.* Because French writers denied their situation in time and failed to articulate their own authentic problem they failed to achieve a writing that is one for all.

> "For us, the historical relativism, posing the *a priori* equivalence of all subjectivities rendered all value to the living event and brought us back

into a literature via an absolute subjectivism to a dogmatic realism. . . . I showed how "retrospective" literature translates amongst its authors as a position of superiority in relation to the whole of society and how those that choose to recount from the point of view of history made, seek to deny their body, historicity, and the irreversibility of time. . . . One doesn't transcend it in escaping, but in assuming it in order to change it."

« Pour nous, le relativisme historique, en posant l'équivalence a priori de toutes les subjectivités, rendait à l'événement vivant toute sa valeur et nous ramenait, en littérature, par le subjectivisme absolu au réalisme dogmatique. . . . J'ai montré comment la littérature « rétrospective » traduit chez ses auteurs une prise de position en survol par rapport à l'ensemble de la société et comment ceux qui choisissent de raconter du point de vue de l'histoire faite cherchent à nier leur corps, leur historicité et l'irréversibilité du temps . . . on ne la transcende pas en fuyant mais en l'assumant pour la changer. » (Situations II, 256)

There are some aspects in this passage, which resonate with themes in *Mille Plateaux* and in *Dialogues.* However in Deleuze's writings terms such as one's historicity, freedom, the writer's obligation, responsibility, the writer's escape, evolve to mean something radically different. These differences that distinguish Deleuze's response to literature from that of Sartre will be addressed in more intricate detail later in this chapter as well as in the following chapters. At present this reading turns to consider the influence that the French philosopher Jean Wahl had on Deleuze's thought. This discussion of Wahl and what he philosophically presents and promotes on the French intellectual scene will provide some insight into those places where Deleuze philosophically diverges from Sartre and other French thinkers.

THE IMPORTANCE OF JEAN WAHL

Jean Wahl, French philosopher, university professor, and poet, precedes both Sartre and Deleuze and mediates the importance of Nietzsche, Kafka, and what Deleuze calls "la pensée américaine," to the French scene as early as 1920. This text asks several questions to consider Jean Wahl's importance: How did Jean Wahl's writings and "teachings" influence the work of both Sartre and Deleuze? In what way and through what venues did Jean Wahl promote Sartre as a thinker? In what way did Wahl act as the precursor to what Foucault names the "*Deleuzian age*"? In other words, how do Wahl's life, writings and teachings facilitate and implement a progressive philosophical movement on the French intellectual scene?

Rhiannon Goldthorpe who writes a critical examination of Sartre's novel, entitled *La Nausée* writes that Sartre suffered a period of stagnation after he left the Ecole Normale and began a teaching post at a lycée in le Havre. Sartre began to feel intellectually reinvigorated when he met Raymond Aaron on vacation from a post at the French Institute in Berlin and began reading Jean Wahl's book *Vers le concret* (1932). Sartre describes the combination of these experiences as "a radical intellectual discovery." (LNC, 30)[4]

> "Jean Wahl dealt primarily with the philosophies of William James, Whitehead and Gabriel Marcel, but related them succinctly to the thinking of Husserl and, particularly, to Heidegger. In Wahl's book Sartre would have found an account of attempts to rethink one of the classical issues of philosophy, the subject-object relationship, an acknowledgement of the limits of analysis *vis à vis* our apprehension of reality, and the view that, although there is no barrier between the mind and things, the mind cannot completely assimilate or exhaust its objects. Nor does the mind 'contain' an immutable, stable self independent of what is not the self. . . . Consciousness is the 'being–in- the-world' of Heidegger, while being itself is simply given: it cannot be accounted for in terms of any pre-existing principles. The recent emphasis in French philosophy on the pure activity of the mind is undermined: the world is a context of instruments and obstacles for man's practical activity, while questions of possibility and impossibility, necessity and contingency are matters of feeling as much as, or more than, of purely intellectual judgment." (LNC, 31)

It is interesting that Jean Wahl's *Vers le Concret* influenced Sartre in his major projects to come, being his novel *La Nausée* and his famous "meteor of a book," *L'Etre et le Néant.* Although Wahl criticizes certain aspects of Sartre's existential writings such as its overvaluation of angst and negativity, his insistence on giving a "signification" to being, and defining the being in-self and for-self as separable entities. Nonetheless, Jean Wahl views Sartre's work as a positive contribution that stimulates a progression in philosophy. Wahl in fact writes several books which discuss Sartre among other existential writers. His book, *Petite histoire de l'existentialisme* (1947), resulted from a conference Wahl held in 1946 at the "Club Now," with notorious French intellectual figures such as Alexandre Koyré and

Emmanuel Levinas attending and participating. Wahl clearly stood apart as the "expert" on the various strains of thought associated with existentialism or a philosophy that centers its questions, problems on the issue of existence or existents. Jean Wahl, although aware of the impasses of existentialism in all of its different forms, whether that of Heidegger, Kierkegaard, Jaspers or Sartre, displays an enthusiasm nonetheless as he describes these philosophical movements as a new stage of philosophizing. A stage that dislodges philosophy from its ontological underpinnings that revolve around such concepts as absolute essence and meaning.

> "Thanks to existentialism, to be or not to be has again become the question. And this reminds us that there have been existentialists- or, as Kierkegaard would say, many existents. We have just intimated that Hamlet was an existent. We could say the same of Pascal, of Lequier, the philosopher from whom Sartre has borrowed the diction: "Faire, et en faisant, se faire" (Do and in doing, make oneself); of Carlyle; and of William James. We could say the same of Socrate's great enemy, Nietzsche. . . . It is clear that one of the consequences of the existentialist movement and the philosophies of existence is that we have to destroy the majority of the ideas of so-called "philosophical common sense," and of what has often been called "the eternal philosophy." In particular, we have to destroy the ideas of Essence and Substance. . . . In this sense, we are witnessing and participating in the beginning of a new mode of philosophizing." (A Short History of Existentialism, 32–24)

In his book, *Les philosophies de l'existence* published ten years later Wahl revisits these same themes although he goes into more depth concerning different aspects of various existential writings. Despite the lapse of a decade Wahl never wavers in his enthusiasm for the ultimately positive, philosophical consequences of the movement that undermines ontology. Wahl restates the same major points being that the existentialist movement makes it necessary to destroy the majority of standard ideas upon which "eternal philosophy" is grounded, for example, its preexisting and rational "essences."

> "One of the consequences of the existentialist movement is that we have to destroy most of the ideas of common philosophical meaning and of that which one often calls eternal philosophy, in particular ideas of essence and substance. Thanks to this movement, we become conscious of the necessity of putting the philosophical concepts into question, to deny the existence of preexisting essences and rationales."

> « Une des conséquences du mouvement existentialiste est que nous avons à détruire la plupart des idées du sens commun philosophique et de ce qu'on a appelé souvent la philosophie éternelle, en particulier les idées d'essence et de substance. Grâce à ce mouvement, nous prenons conscience de la nécessité de remettre en question les concepts philosophiques, de nier l'existence d'essences préexistantes et rationnelles. » (Philosophies de l'existence, 160)

Although Jean Wahl brings up many problems inherent to existentialism, in its defense he urges that regardless of its failures one should esteem it for challenging the foundations of "Eternal Philosophy." At the 1946 conference held by the "Club Now," George Gurvitch criticizes what he perceives as the vicious circle aspect of the existentialist movement and turns to Jean Wahl's refusal to be labeled "existentialist" as substantiating his own position. However, despite Wahl's awareness of the problems inherent in existentialism his position remains more fluid and comprehensive as he suggests that existentialism's failings are integral to its value. Gurvitch states,

> "First of all, I would like to congratulate Jean Wahl on having been able to say *No* to the student who asked him if he was an existentialist. I would even like to hope that this No will eventually grow into complete non-acceptance. . . . Sartre's *L'Etre et le Néant* proclaims a possible liaison between the logomachy of Hegel and the philosophy of existence. To become "existentialism," existence first passes through the logonomical purgatory of "in-itself" and "for-itself" to rediscover itself—impoverished to the limit. If one could accept the opening chapters of the work of Sartre, I believe one could just as easily, and far more sensibly, accept purely and simply Hegel or dialectical materialism. . . . One affirms existence after one has carefully emptied it of all its richness, all its contradictions, all its collective and historical aspects. . . . As the traditional empiricism amounted to a total destruction or transformation of experience into chaos of sensation, so existentialism applies itself to the task of reducing existence to zero. This is the nausea of impotence." (A Short History of Existentialism, 38, 39)

This criticism does bring up certain problematics which Wahl also addresses in a question form: "There is, however, a question which may trouble the mind, and even the existence, of the existentialist. Does he not risk destroying the very existence which he wishes above all to preserve . . . ? Is it for the existent to say that he exists? In short, is it, perhaps, necessary to choose between existentialism and existence? Such is the dilemma of existentialism." (*A Short History of Existentialism,* 39). However, despite this Jean Wahl disagrees with the statements of Gurvitch. "I could not let pass, without registering a protest, the remarks of my friend Gurvitch whose concluding words were perhaps more forceful than premeditated."(39).

Jean Wahl counters another criticism voiced by Nicolas Berdiaeff eventually in his book *Les Philosophies de l'existence.* Berdiaeff brings up the issue that Sartre and Heidegger strove to promote their own sort of ontology that holds an inherent contradiction if what they claim to reveal are the meanings of being-in-the world. For any ontological approach necessarily would as Gurvitch said, empty existence of its true complexity and richness. Berdiaeff says,

> "Ontology is impossible from the existential point of view. Jaspers was certainly nearer the truth when he said that the only possibility would be a perusal of ciphers, a symbolic knowledge far removed from any rational ontology. Yet Heidegger and Sartre want to create a rational ontology, Sartre even more than Heidegger. . . . Why is an ontology impossible? Because it is always a knowledge objectifying existence. . . . So that in ontology—in every ontology—existence vanishes. There is no existence because existence cannot be objectified. . . . It is only in subjectivity that one may know existence, not in objectivity. In my opinion, the central idea has vanished in the ontology of Heidegger and Sartre." (A Short History of Existentialism, 36, 37).

Jean Wahl replies that the contradictions and ambiguities in these various forms of existential philosophy must be seen rather as a value in that these shortcomings reaffirm the *échec* of ontology. Wahl also asserts that these "ontologies manquées" think of themselves as such. The title of Heidegger's book, "Holwege" attests to this as it signifies, "Ways Leading Nowhere" or "Lost Tracks." Because of the diversity of "existential" philosophies, which all bear a multiplicity of variant beings, they as an ensemble demonstrate that there indeed can be no one, unified ontology that can explain the "reality," and "truth" of the infinite variations of existence. Therefore the movement makes the failure of reason unquestionable and negates ontology.

> "Therefore we are before ontologies, but we could say unsuccessful ontologies, which expressly think of themselves as such. This is perhaps besides the signification of the title of Heidegger's last book: Holwege, which means something like "Paths that lead nowhere," "Lost ways" . . .

by the fact that, in Heidegger, Jaspers, Sartre, it is multiple . . . this ontology is a defeat for reason . . . we could say that it is a failure for ontology itself; because can one yet call ontology the affirmation of multiples of categories of being? The idea of ontology in the plural is this not the negation of ontology itself?"

« Donc nous sommes devant des ontologies, mais nous pourrions dire des ontologies manquées, qui se pensent manquées forcement. C'est d'ailleurs peut-être la signification du titre du dernier livre de Heidegger: Holwege, qui signifie quelque chose comme « Voies qui ne mènent nulle part », « Voies perdues » . . . par le fait que, chez Heidegger, chez Jaspers, chez Sartre, il est multiple. . . . Cet ontologique est un échec pour la raison . . . nous pourrions dire qu'il est un échec pour l'ontologie lui-même; car peut-on encore appeler ontologie l'affirmation de multiples catégories d'êtres? L'idée de l'ontologie au pluriel n'est-elle pas la négation de l'ontologie elle-même? » (Philosophies de l'existence, 64, 65).

For Wahl, the failures of each of the variations of existential thought paradoxically render it a success as a movement. In addition even if Wahl criticizes Sartre's definition of "being" and his pour-soi and en-soi, he praises Sartre for exemplifying one of the main points of Wahl's premise which affirms; in that being constantly becomes, what being reveals or discovers will never be equitable, stable, congruent or unified.

"All that we have said shows the difficulty that there is in judging these philosophies in a general way, first due to their diversity, second, even in the case of one author, and particularly, with Sartre, because of the diversity of solutions following the periods of his thought. . . . But this ambiguity is one of the reasons perhaps for the value and success of these philosophies, because it characterizes the man of our times. . . . Sartre's thought expressed in *La Nausée* is very different from that which he expresses in *Les Mouches.* We have seen a duality between his idealism and realism, between his ontological and phenomenological tendencies, between the idea of freedom and engagement, between a stoicism which sometimes appears and a recommendation for action, between pessimism and confidence. . . . Impossibility of any justification, and search for it, the two motifs are mixed up in Sartre's philosophy."

« Tout ce que nous avons dit montre la difficulté qu'il y a à juger d'une façon générale ces philosophies, d'abord à cause de leurs diversités, ensuite, pour un même auteur, et, particulièrement, pour Sartre, à cause de la diversité des solutions suivant les époques de sa pensée. . . . Mais cette ambiguïté est une des raisons peut-être de la valeur et aussi du succès de ces philosophies, car elle caractérise l'homme de notre temps. . . . La pensée de Sartre telle qu'il exprime dans la Nausée est très différente de celle qu'il exprime dans Les Mouches. Nous avons vu de même une dualité entre son idéalisme et son réalisme, entre ses tendances ontologiques, ses tendances phénoménistes, ses tendances phénoménologiques, entre l'idée de liberté et l'idée de l'engagement, entre un stoïcisme qui apparaît parfois et une recommandation de l'action, entre le pessimisme et la confiance. . . . Impossibilité de la justification, recherche de la justification, les deux motifs sont mêlés dans la philosophie de Sartre. » (Philosophies de l'existence 161).

Wahl stresses again in the same book that the greatest value of Sartre's work; what gives it force and influence resides in the innumerable discrepancies that one finds throughout whether in his fiction or philosophical writings. Sartre himself seemed to call attention to this very aspect of his "existence": "I am a walking contradiction." (LNC, 24) [5]

"The very ambiguities of existentialist thought, at least as they are seen in that of Sartre, have contributed to the power of its influence. That which was presented first in *La Nausée* and even in *Being and Nothingness* as a pessimistic philosophy, appeared in *Les Mouches,* for example, or in certain political studies by Sartre, as a doctrine of hope." (Philosophies de l'existence, 135)

« Les ambiguïtés mêmes de la pensée existentialiste, telles du moins qu'elles se voient chez Sartre, ont contribué à la puissance de son influence. Ce qui se présentait d'abord dans *La Nausée* et même dans *L'Etre et le Néant* comme une philosophie pessimiste est apparue avec *Les Mouches,* par exemple, ou avec certaines études politiques de Sartre, comme une doctrine d'espoir. » (Philosophies de l'existence, 135)

Jean Wahl who taught, influenced Sartre, and eventually promoted Sartre on the French scene had an impact on Deleuze as well. In fact as already mentioned Deleuze singles Wahl out as one of the two "most important" French philosophers. « A part Sartre . . . le philosophe le plus important en France, c'était Jean Wahl. » It seems almost suspect that Deleuze names Jean Wahl one of two of France's most *important* philosophers. Why?

Because he is certainly not a focus of scholarly attention in the way that Blanchot, Lacan, Foucault, and Bataille are. One can scarcely find his publications or others that refer to his works or life in major research libraries. Several questions will be addressed to clarify the role Jean Wahl played in Deleuze's evolution as a thinker. These questions include; where did Deleuze come into contact with Jean Wahl? What aspects of Wahl's writings might have resonated most deeply with Deleuze? What aspects of the "pensée américaine" that Wahl introduced to Deleuze and his contemporaries held so much importance?

Despite the fact that the work of Jean Wahl is not widely discussed in the United States does not discount the fact that he was indeed an extremely important philosopher and *artist*-thinker as Deleuze states. For Wahl collaborated with great painters like Marc Chagall, and taught many famous French intellectuals and literary artists while he was a professor of philosophy and literature at the Sorbonne from 1936 to 1967. Jean Wahl's own scholarly and literary contributions spanned a period of more than five decades. Jean Wahl, like Sartre, was an influential philosopher who Deleuze encountered personally. He not only had an intellectual impact on Deleuze but surely had an affective impact as well. That is to say that Wahl as did Sartre in some way mentored Deleuze in his formative years as a thinker and Deleuze felt great respect for him as his words of homage make clear. Deleuze attended classes taught by Jean Wahl who taught philosophy at the Sorbonne in the same years Deleuze attended, (1944–1948), culminating in Deleuze passing the Agrégation in1948. In that same year Jean Wahl published *The Philosopher's Way* and dedicated it to his students at the Sorbonne. Considering the affinity between Wahl and Deleuze one can assume that Deleuze was aware of this professor's particular attention to his students and sincerity in wanting to form them to take up "The Philosopher's Way." This book is interesting for diverse reasons. First because it attests to the fact that Wahl helped mediate Sartre's philosophy not only to his fellow "professional" intellectuals but to his students as well. Wahl also wrote the book lamenting the fact that this "textbook" would fail to be "revolutionary" due to the contradiction in terms. However Wahl calls for a philosophical "revolution," for the inception of a new mode of philosophizing. The book reveals one of the sparks that may have ignited Deleuze's intellectual fire. In the book Wahl writes nineteen chapters that give a comparative overview of all major philosophical figures and their different or overlapping responses, interpretations and applications of various concepts such as: substance, being, existence, reality, becoming, freedom, etc. Jean Wahl ends his book as if advising; know these thinkers, then bid them farewell. Venture beyond exhausted, systematic approaches that rely on absolute truths and logic based concepts. Create works that incorporate the values of aesthetics.

> "This history of philosophical thought, like that of humanity at large, has been a glorious but unhappy one. Remaining aware of the achievements of great thinkers, we must look elsewhere for a richer more adequate view of reality. . . . We are on the eve of a revolution of thought analogous to that which took place when man turned from the ancient world with its diffidence toward the infinite, toward time, and toward matter. . . . Now we have to give a new, less conceptual, form to these ideas of infinity, time, and matter. Perhaps this revolution will be even greater that the preceding one. Its dangers and the possibilities of its misinterpretation (already illustrated in some elements of Nietzsche's thought, and particularly as they have been misconstrued by some of his apparent and superficial followers) must not blind us to the necessity for it." (The Philosopher's Way, xi-xiii)

In this introductory passage in *The Philosopher's Way,* Jean Wahl emphasizes that what is of paramount importance for philosophy is not the variety of approaches to the plethora of common ontological concepts that the book plots out; but the revolution on the horizon that will sweep away these concepts and offer a richer less conceptual view of reality. As Wahl dedicates the book to his students at the Sorbonne, he obviously calls them to activate this new movement in philosophy. Or perhaps more accurately he urges them to transform philosophy as it has operated and been known up to date. Jean Wahl singles Nietzsche's work out as an example of the most recent philosophical revolution. Jean Wahl's remarks about Nietzsche are themselves revolutionary: First because Nietzsche was not part of the traditional philosophical cannon taught in the university. Academics at this time labeled Nietzsche a philosopher-poet rather than a model of serious scholarship. Secondly, at this moment in history, in the wake of World War II, just one year after the defeat of Nazi Germany, the popular view still associated Nietzsche's thought with Hitler's murderous, fascist regime. In addition, Wahl's defense and promotion of Nietzsche is that much more extraordinary as he himself, a French Jew, had to seek exile in the United States after officials released him from the Drancy concentration camp. Jean Wahl vindicates Nietzsche so that eventually others may seriously consider his work as a philosopher and further revolutions will follow. Wahl states that contrary to popular thought and propaganda, Nietzsche's thought was neither political nor fascist but rather victimized by *superficial followers* who misinterpreted and propagated it. (*Philosopher's Way,* xiii). Wahl, one of the most respected "connoisseurs" of the history of philosophy, essentially set the stage for a large scale integration of Nietzsche into French philosophy, but this movement didn't find its full force until two decades later.[6]

Deleuze who became one of the foremost French writers on and through Nietzsche from the sixties on, praises Wahl who obviously provided the same *Courant d'air* as did Jean Paul Sartre into an intellectual atmosphere that Deleuze found stifling. Surely Deleuze who describes feeling bludgeoned to death with the history of philosophy, feels indebted to a professor who emphatically announces in a "textbook" that the most important matter for young philosophers of the present and future, is not that they simply swallow all that has been said and done up to date in philosophy as written in textbooks and explicated by professors. The young philosophers must know history only to recognize what Wahl's book documents; that traditional philosophy and its concepts have gotten to a stage of exhaustion and reach nothing more than impasses. The new generation of philosophers must create new, less conceptual "concepts." They must take all that has gone before and in a spirit akin to Nietzsche's, transform every conceptual '*it was*' into a scintillating '*thus I willed it.*' (Ecce Homo, 309). In other words, any particles of value must be reinvigorated and translated into forms, styles that render new vitality and meaning. These forms or styles will move increasingly away from a dialectical philosophy and closer to an aesthetic-philosophy.

What does this mean, an aesthetic-philosophy? Perhaps to refer to two other passages in Wahl's book may help elucidate this idea. On page three in the first chapter entitled, "The Necessary Revision of Metaphysical Concepts," Jean Wahl suggests that philosophy needs to revolutionize not only its concepts but also its procedures. "The very fact that from antiquity to the present day a kind of revolution has been going on in some of these ideas will incline us to the belief that a similar revolution may take place now." In the conclusion, 320 pages later, Wahl tells the reader that this revolution will question the very foundations of philosophy articulated in founding concepts such as "being," and "cause." As Jean Wahl continues, what he communicates becomes increasingly elusive. He articulates that although philosophy's quest is for knowledge, this knowledge can not necessarily be grasped or "reduced" through "intellectual understanding." He redefines philosophy as a process that perpetuates a particular form of questioning that renders a particular movement. This movement articulates problems, rather than propagates solutions or interpretations. It does not translate irreducible realities into absolute "truths." This questioning in addition follows a movement described in cryptic terms. *This movement cannot be seen but only dimly discerned from reality through dialectics and the antithesis toward ecstasy. (Philosopher's Way, 324).* Wahl's language becomes increasingly vague as it speaks of the realm of art and aesthetics as one that *achieves worlds,* and transmits experiences of fullness, void of arbitrary separations, based on binary systems that polarize inner/outer, infinite/finite, philosophy/art: In this realm all concludes in silence. (Philosopher's Way, 324).

> "We shall not even say that philosophy is a science in the ordinary sense of the word. It is necessarily a quest, a search for knowledge, but for a knowledge that is not necessarily reducible to intellectual understanding. The goal of philosophy might be more akin to what Alexander has called 'compresence with things. . . . Questions . . . are philosophy itself, for philosophy is rather questioning than answering. It is movement, dimly discerned, rather than seen, from reality through dialectics and the antithesis toward ecstasy. In the presence of works of art, or achieved worlds, or simply things, we experience a fullness of being, we no longer separate the inner and the outer, the infinite and the finite, and the unceasing dialogue comes to its conclusion, in silence." (Philosopher's Way, 3, 324)

After 320 some pages of the history of philosophical concepts, the words in the quotation just above conclude Jean Wahl's "textbook." One asks what Wahl suggests by all of this. After his very thorough outline of varied philosophical paradigms, which he articulates in a traditional fashion, Wahl seems to conclude that these paradigms respectably attempt to describe man's existence and relationship to the world and its reality, but they do not go far enough. Philosophy needs to generate new methods and less material concepts that relay sensory and other forms of non-intellectual "knowledge." Perhaps one of Wahl's greatest contributions to the French philosophical tradition lies in his own revolutionary stance that challenges the primacy of Being, calls for anti-ontologies, and pays heed to those "philosophically" unclassifiable thinkers that most traditional French philosophy programs and text books leave out, thinkers such as: Bergson, Kierkegaard, Holderlin, Proust, James, Heraclites and Nietzsche. « Il y a un devenir-philosophe qui n'a rien à voir avec l'histoire de la philosophie, et qui passe plutôt par ceux que l'histoire de la philosophie n'arrive pas à classer. » "There is a becoming-philosopher which has nothing to do with the history of philosophy, and which passes rather through those that the history of philosophy can't manage to classify." (Dialogues, 8).

In his juxtaposition of Jean Paul Sartre and Jean Wahl in *Dialogues* (expressed in the opening citation of this chapter), Deleuze underlines that Jean Wahl's ability to challenge the primacy of the verb "Etre": In other words the unquestioned authoritarian status of ontology, is what makes him so "important." Deleuze goes further to state that in Wahl's work this is an active challenge as Wahl stretches the French language with what Deleuze names *the art of the ET.* Through this *art* Jean Wahl achieves what is almost impossible in a language policed by institutions, such as the "Académie Française" that abhor linguistic mutations. Deleuze explains that Wahl pushed the French language to a place that allowed his listeners to think entirely "new" things within and with it. Whereas Sartre philosophically speaking gets trapped in the verb "to be," in "Being," in ontology, Jean Wahl slips outside and destabilizes its structures with the conjunctive force of the "ET." (Dialogues, 72). Again in a note in *Mille Plateaux* Deleuze refers to this same exemplary quality in Jean Wahl's work. "It should not be thought adequate to analyze the "and" as a conjunction; rather, "and" is a special form of every possible conjunction and brings into play a logic of language. Jean Wahl's works contain profound reflections on this sense of "and," on the way it challenges the primacy of the verb "to be."" (Mille Plateaux, 526).

In the section of *Mille Plateaux* that leads to this note Gilles Deleuze goes into detail on the function and importance of the "and" and its particular status in American literature.

> "There has always been a struggle in language between the verb *être* (to be) and the conjunction *et (and).* It is only in appearance that these two terms are in accord and combine, for the first acts in language as a constant and forms the diatonic scale of language, while the second places everything in variation, constituting the lines of a generalized chromaticism. From one to the other everything shifts. Writers in British or American English have been more conscious than the French of this struggle and the stakes involved, and of the valence of the "and." . . . To be a foreigner, but in one's own tongue. . . . To be bilingual, multilingual, but in one and the same language, without even a dialect or patois." (Thousand Plateaus, 98)

In that the ET and the verb ETRE always struggle within the domains of language and thought, and Sartre remains stuck in the trappings of the verb ETRE whereas Wahl writes and philosophizes as the exemplary French practitioner of the "method" of the ET, (Dialogues, 72), it would seem that although Deleuze describes Sartre as his "Master" at one point, Deleuze more closely aligns his own philosophical practice to Wahl's.[7]

One might argue that one of the missing elements in Sartre as opposed to Jean Wahl is Nietzsche. Wahl articulates at a 1964 Nietzsche conference that ontological aspects are absent in Nietzsche due to the fact that he does not believe in the "concept" Being. (Cahier, 225). In contrast, Sartre's reluctance to shed himself of the influence of Hegel, Heidegger, and Husserl limits his ideas and retains them in an ontological field of meaning and reference that depends on a "defined" Being, a conceptualization of Being. Wahl in fact most explicitly criticizes Sartre for his attempts to define "Being" in Sartre's, *Etre et le Néant.* Wahl in fact underlines that Sartre's conceptualization of 'Being' stands out as the work's primary shortcoming.

> "The effort to give significance to the term 'Being' seems bound to fail. Sartre divides Being into two kinds—Being in itself and Being for itself. But this distinction is not quite satisfying, since what Sartre describes as Being in itself is purely an abstraction; he means by this expression the separate things in their separateness, for example a table, a wall, the root of a tree. But there are no things in themselves in the sense which Sartre uses this expression. . . . As for the Being for itself, by which Sartre means consciousness, he defines it by its relation to Nothingness, which would lead us to think that one of the kinds of Being is a kind of Not-Being. . . . Moreover, Sartre has to relate these two kinds of Being. According to him, Being for itself is the product of a kind of subtraction or even annihilation performed within Being in itself, and we are introduced here to a nearly cabalistic conception." (The Philosopher's Way, 54).

In *Dialogues,* Deleuze writes that "German" or German philosophy is haunted with the importance of and nostalgia for "être"/being that is associated with all the rooting, tree structures. Those that break free, break out of this "nostalgia," "foreigners in their own language," i.e. Nietzsche, Kafka, American-Blacks do so with the "art" of the Et, which makes language move to destabilize and unground the roots and static "being" formulas of the insular, inside *Grund* system. Deleuze states that the "art of the ET/AND" is not about speaking a language as a foreigner but concerns being a foreigner in one's own language to "make it move." Deleuze explicates that it is in this sense that the American language is the Black's language. On the other hand it is almost impossible to make German "move" due to the fact that German is haunted by the primacy of being and makes all the conjunctions which it uses to create a composite word tend towards it. German remains faithful to "the cult of the *Grund,*" to the tree and root structures, to the "Inside." In opposition English creates composite words with an implied AND, that ever relates to the Outside, it is never grounded, and is free of foundations. English runs on the surface as the rhizome. "AND . . . AND . . . AND . . . Empiricism is nothing other than this." (Dialogues II, 59, 32).

Symbolically, this passage in *Dialogues* places Jean Wahl on the side of the *ET*, "*black*" Americans, "empiricism" and Nietzsche. As for Sartre, despite his "richness" and originality, he gets caught on the German side, to that which is root bound, to the legacy of (Husserl, and Heidegger). Deleuze tells us that when one introduces the "Et créateur" the major language breaks down and is forced to flee. Deleuze concludes this passage to define the process again in terms of "empiricism and pragmatics." "This is empiricism, syntax and experimentation, syntactic and pragmatic, the case of speed." (Dialogues, 73). Strangely enough, or coincidentally, more than three decades prior, George Bataille like Gilles Deleuze, speaks of the value of Jean Wahl while he juxtaposes him to Sartre. In September 1946 George Bataille renders this homage to Jean Wahl in the revue *Critique.* Bataille also, as does Deleuze, places Sartre on the side of the Germans and Wahl on the side of Nietzsche. Furthermore Bataille also refers to the pragmatic dimension of Wahl's thought as that which grants it an ungraspable and fluid quality.

> "More discretely than that by Jean-Paul Sartre, the lively/living philosophy in France is represented by Jean Wahl. Wahl has the advantage over Sartre due to an almost incomparable mastery of the knowledge of the history of philosophies, but he is not any less of an original philosopher, he is even a "philosopher-poet," rather far from the professorial tradition. [Here evoking *Human Existence and Transcendence,* 1944, and *Kierkegaardian Studies,* 1938.] If the word hadn't taken on such a confused meaning in the mind of the public, this thought could be qualified as existential: But it is especially and voluntarily fluid, ungraspable, being pragmatic. Its profound difference with that of Sartre could be perhaps expressed in a fundamental way if one distinguishes, in the existential tradition, on the one hand Kierkegaard and Nietzsche and on the other the professor-philosophers of modern Germany (Husserl, Heidegger). Wahl is especially preoccupied with Nietzsche and Kierkegaard, Sartre by the phenomenologists." (Nietzsche en France, 184).

> « Plus discrètement que par Jean-Paul Sartre, la philosophie vivante en France est représentée par Jean Wahl. Wahl a sur Sartre l'avantage d'une maîtrise presque incomparable dans la connaissance de l'histoire des

philosophies, mais il n'est pas moins un philosophe original, même un « philosophe-poète, » assez loin de la tradition professorale. [Sont évoqués Existence humaine et transcendance, 1944, et Etudes kierkegaardiennes, 1938.] Si le mot n'avait pris dans l'esprit du public un sens très confus, cette pensée pourrait être qualifiée d'existentielle: Mais elle est surtout et volontairement fluide, insaisissable, étant pragmatique. Sa différence profonde avec celle de Sartre est peut-être exprimée d'une façon fondamentale si l'on distingue dans la tradition existentielle, d'une part Kierkegaard et Nietzsche, et de l'autre les philosophes-professeurs de l'Allemagne moderne (Husserl, Heidegger). Wahl est préoccupé surtout par Nietzsche et Kierkegaard, Sartre par les phénoménologues. » (Nietzsche en France, 184).

It would seem that Bataille's reasons for esteeming Wahl mirror those expressed by Deleuze. As stated in Bataille's praise, Jean Wahl moved in both his writings and movements as a "philosopher-poet," someone that could not be positioned in any one system, category or group. Wahl like Sartre also lived and worked, as a "walking contradiction," as anti-system and anti-Professor professional, while at the same time was a living icon of the French University. In fact, Wahl's distrust and defiance of categories prompts him to deny being an "existentialist." "One day not long ago, as I was leaving a café in Paris, I passed a group of students, one of whom stepped up to me and said; "Sûrement, Monsieur est existentialiste." I denied that I was an existentialist. Why? I had not stopped to consider, but doubtless I felt the terms suffixed by *ist* usually conceal vague generalizations." (Short History of Existentialism, 1).

As Jacques Le Rider points out in his book *Nietzsche en France,* Wahl's rareness as a university philosopher stemmed from his anti-traditional interests and selection of "associates": Wahl frequented Georges Bataille and Roger Caillois, and participated in the meetings of *le Collège de Sociologie* from its inception in 1937 up until 1944 when Wahl had to exile to the United States after his release from the concentration camp in Drancy, France.[8] As already mentioned, Wahl also was one of the first French intellectuals to consider and present Nietzsche's work as seriously as that of Plato or Hegel. This transmission of Nietzsche on the French scene in fact spanned a period of over thirty years. Jacques Le Rider states,

"Born in 1888, named to the Sorbonne in 1945, this normalian (*From the Ecole Normale Supérieure*) was one of those rare university philosophers to maintain a relationship with Georges Bataille or Roger Caillois, one of the first to treat Nietzsche with the same seriousness as Hegel or Kierkegaard. . . . He played an important role in the transmission of

> Nietzscheism from the thirties to after the war. . . . Having published two polycopied courses on Nietzsche in 1959 and 1961, Jean Wahl prefaced a new collection of the anthologized edition of Nietzsche edited by Henri Albert." (Nietzsche en France, 183–185)
>
> « Né en 1888, nommé à la Sorbonne en 1945, cet ancien normalien fut un des rares philosophes universitaires à entretenir des relations avec Georges Bataille ou Roger Caillois, un des premiers à traiter de Nietzsche avec le même sérieux que de Hegel ou de Kierkegaard. . . . Il joua un rôle important dans la transmission du nietzschéisme des années 30 à l'après-guerre. . . . Après avoir publié deux cours polycopiés sur Nietzsche en 1959 et 1961, Jean Wahl préface en 1963 une nouvelle édition de l'anthologie de Nietzsche éditée par Henri Albert. » (Nietzsche en France, 183–185).

What Nietzsche offers for both Wahl and Deleuze is an alternative to negativity, pessimism and a structure that names lack the primary force that creates desire, and stimulates action.[9] Wahl also traces the preeminent idea of liberty in "existential philosophies" to Nietzsche's thought. Wahl praises Nietzsche for emphasizing the force of will and freedom that together make it possible to create new worlds with new values.

> "No theory, a part from that of Nietzsche, had put the idea of man's creation of values in such full light. Moreover, it is certain that the place given to the idea of freedom and the affirmation that freedom always persists in us had been due to something in the resonance of these philosophies."
>
> « Aucune théorie, sauf celle de Nietzsche, n'avait mis en aussi pleine lumière l'idée de la création des valeurs par l'homme. En outre, il est certain que la place faite à l'idée de liberté et à l'affirmation que la liberté persiste toujours en nous a été pour quelque chose dans le retentissement de ces philosophies. » (Philosophies de l'existence, 135).

Despite the fact that freedom and man's role in creating values in the world became a common and fundamental aspect in existentialism, Wahl emphasizes that Sartre's idea of *freedom,* that stems from a Hegelian, dialectic sense of negativity, indeed negates itself. Jean Wahl goes on to explicate that in such a model, freedom can only become associated with a sense of limitation and becomes an essentially un-free "freedom."

> "Freedom always appears in Sartre, Jaspers and Heidegger, as limitation and finitude. . . . Sartre goes even further: freedom is, in a sense, a means of being, it is a deficiency, a lack, a sort of hole, of nothingness in the interior of being, and here, this is all the theory of for-self, in Sartre . . . I am being which I am not and I am not the being that I am. Freedom will be linked to my essential negativity (in the Hegelian and dialectical sense, and also in the satrtrian meaning of the word.)."
>
> « La liberté apparaît toujours, chez Sartre, d'ailleurs comme chez Jaspers et Heidegger, comme limitation et finitude. . . . Sartre va même plus loin : la liberté est, en un sens, un moindre être, elle est une déficience, un manque, une sorte de trou, de néant à l'intérieur de l'être, et, ici, c'est toute la théorie de pour-soi, chez Sartre . . . Je suis l'être que je ne suis pas et je ne suis pas l'être que je suis. La liberté sera liée à mon essentielle négativité (au sens hégélien et dialectique, et aussi au sens sartrien du mot.) » (Philosophies de l'existence, 91).

Wahl discusses this problem of negativity in Sartre's idea of freedom when he underlines that Sartre's individual is never free to *not be* free, and that this "freedom" is embedded in a certain situational context of time, place and politics, what motivates the individual to act, negates the term "freedom." In a genteel fashion, Wahl praises Sartre for insisting on the idea of a never-wavering state of freedom despite the fact that Sartre's freedom fails.

> "We are not free to not choose, we are not free to not be free . . . this indicates that there is a problem at the interior to Sartre's thought–. . . . For in Sartre's philosophy we find ourselves before two problems: In the first place, that of situation. In effect, freedom is always a situational, limited, conditioned freedom. Therefore, each of our actions can be interpreted in a double way: in function of the situation or in function of freedom. . . . In the end, our freedom tends to extinguish itself. "
>
> « Nous ne sommes pas libres de ne pas choisir, nous ne sommes pas libres de ne pas être libres . . . et cela indiquerait qu'il a y là un problème à l'intérieur de la pensée de Sartre–. . . . Pour la philosophie de Sartre, nous nous trouverions en face au moins de deux problèmes. En premier lieu, le problème que pose l'idée de situation. En effet, la liberté est toujours liberté en situation, liberté limitée, conditionnée. Ainsi, chacune de nos actions peut être interprétée d'une double façon : en fonction de

la situation ou en fonction de la liberté. . . . Enfin, notre liberté tend à se nier elle-même. » (Philosophies de l'existence, 90, 91, 97).

Examples of the problematics inherent in Sartre's idea of "freedom" can be found in many passages in Sartre's publication, *Situations II.*

"Writing, this is neither life, nor to run from life in order to contemplate platonic essences and archetypes of beauty in a world at rest. . . . This is to exercise a trade. A trade that demands an apprenticeship sustained work, professional conscience, and a sense of responsibilities. Society is the one to put our duties and projects, on our backs."

« Ecrire, ce n'est pas vivre, ni non plus s'arracher à la vie pour contempler dans un monde en repos. . . . C'est exercer un métier. Un métier qui exige un apprentissage, un travail soutenu, de la conscience professionnelle et le sens des responsabilités. Nos charges et nos devoirs, c'est la société qui vient de nous les mettre sur le dos. » (Situations II, 260).

This quote substantiates Wahl's observation that Sartre's sense of freedom, which he ties to responsibility and social and artistic duty, is actually the antithesis to freedom. It is rather as Wahl states, a limited, conditioned choice within a sphere that places duties on the individual's shoulders and even dictates what developmental course a writer must follow. "It is our nature to be free, but one knows that Sartre doesn't admit any "true nature" properly speaking and freedom is not, in effect, a nature in the sense that it is always a possibility for us, to be other than that which we are." « C'est notre nature d'être libres, mais on sait que Sartre n'admet pas de nature à proprement parler et la liberté n'est pas, en effet, une nature en ce sens qu'elle est toujours la possibilité, pour nous, d'être autre chose que ce que nous sommes. » (Philosophies de l'existence, 90)

In the Chapter, *Qu'est-ce que la littérature,* Sartre writes that because the writer exists as a public figure, he/she must then write in accordance with certain preconceived notions that society has of him/her in relation to origins, ethnicity, race, class etc. In fact for Sartre, with his emphasis on the importance of the "other" in the mediation of the individual's sense of identity, the individual's creative projects as an extension of the individual are also subject to respond, to exist, to come into being only in relation to, in consideration of the "other," or the "others" being the society at large. Only if the writer recognizes and assumes this position will he reach the status of the "écrivain engagé."

"I would say that a writer is engaged while he strives to assume the most lucid and entire conscience to be embarked . . . if it is true that it is necessary to demand something of his work relating to his condition, it is necessary then to remember that his condition is not only that of a man but also of a writer. . . . Because there are qualities that come to us uniquely through judgments of others. . . . It is because I become a man that other men consider writer that entails responding to a certain social function. . . . Take the case of the great black writer Richard Wright. . . . Can one suppose that for an instant he accepts to spend his life in contemplation of the True, the Beautiful and eternal Good, when 90 percent of Negroes in the South are practically deprived of the right to vote?. . . . Therefore if a Black in the United States discovers a vocation to write, he discovers his subject simultaneously."

« Je dirai qu'un écrivain est engagé lorsqu'il tache à prendre la conscience la plus lucide et la plus entière d'être embarqué. . . . Seulement s'il est vrai qu'il faut demander des comptes à son oeuvre à partir de sa condition, il faut se rappeler aussi que sa condition n'est pas seulement celle d'un homme en général mais précisément aussi d'un écrivain. . . . Car il y a des qualités qui nous viennent uniquement par les jugements d'autrui . . . c'est que je deviens un homme que les autres hommes considèrent comme écrivain c'est-à-dire qui doit répondre à une certaine demande et fonction sociale. . . . Prenons le cas du grand écrivain noir Richard Wright. . . . Peut-on supposer un instant qu'il accepte de passer sa vie dans la contemplation du Vrai, du Beau et du Bien éternel, quand 90 pourcent des nègres du Sud sont pratiquement privés du droit de vote?. . . . Si donc un Noir des Etats-Unis se découvre une vocation d'écrivain, il découvre en même temps son sujet. » (Situations II, 124).

In this same passage Sartre refers to Kafka in terms of his religious, ethnic and class background: "He is Jewish perhaps, and Czech and of a peasant family, but he is also a Jewish, Czech, country bumpkin writer." « *Il est Juif peut-être, et Tchèque et de famille paysanne, mais c'est un écrivain juif, un écrivain tchèque et de souche rurale.* » Sartre suggests that because a Jew's *Jewness* stems from others considering him as Jew, he must write from the situation that accordingly defines him.[10] Interestingly, Deleuze employs these same adjectives when he speaks of Kafka, but contrary to Sartre, Deleuze stresses that Kafka and all other "*minoritary* writers" that enter and activate *minoritarian-becomings* do so by becoming non-writers and everything and everyone but what society defines and perceives them as *Being.*

The minor writer must betray the order that fixes individuals in categories and must find and extend "lignes de fuites," ways out of the system. "To write is to trace lines of flight which are not imaginary. . . . To write is to become, but has nothing to do with becoming a writer. That is to become something else." Deleuze goes on to convey that the main reason for writing is to betray one's own reign, sex, class and majority. Many people attempt to be traitors but fail because this is the ultimate creative challenge. The subject/ego has to disappear, one must become unrecognizable. Almost as a direct rebuttal to Sartre's prescription that the writer accept his/her responsibility which is tied to his/her identity and social position, Deleuze forwards that "lines of flight," extended through the "traitor's" writing produce the *real,* and generate life while finding and activating new weapons: This active movement is heroic rather than a cowardly escape into the imaginary, or into art. (Dialogues, 54, 56)

Sartre insists that writing entails practicing a trade whereas Deleuze describes writing as tracing lines of "escape," "leakage." Whereas Sartre's writer must conform to the outside world's expectations of what he will write about in terms of his situation, Deleuze's writer must lose his/her identity, his/her face, must disappear and become imperceptible. Whereas Sartre speaks of the obligation of black writers and Jewish writers, like Richard Wright and Kafka, to write for the uplift of *their people,* Deleuze writes that every "minor writing," regardless of the ethnic, racial, national, origins or gender of the writer generates "les devenirs minoritaires." These devenirs liberate all people from the kind of systems that segment individuals into groupings such as, "Negro," "Jew," "Catholic," "Female," "Gay," "Mexican," and at the same time these becomings efficiently yet *minutiously* eat away at the underpinnings of these structures built on static "Being" notions and binary machines.

> "It is not about freedom in opposition to submission, but only about a line of flight/ leakage, or rather a simple *way out, "to the right, left, any possible direction,"* the least significant possible. . . . A writer is not a man-writer, this is a political-man, and this is a machine-man, and this is an experimental-man. . . . There is no subject, there are only collective "agencements" of the enunciation-. . . . The literary machine plays the part of the relay of a future revolutionary machine, not for ideological reasons, but because it only, is determined to fulfill the conditions of a collective enunciation."

> « Il ne s'agit pas de liberté par opposition à soumission, mais seulement d'une ligne de fuite, ou plutôt une simple *issue, « à droite, à gauche, où*

> *que ce fut,* » la moins signifiante possible. Un écrivain n'est pas un homme écrivain, c'est un homme politique, et c'est un homme machine, et c'est un homme expérimental. . . . Il n'y a pas de sujet, il n'y a que des agencements collectifs d'énonciation-. . . . La machine littéraire prend ainsi le relais d'une machine révolutionnaire à venir, non pas du tout pour des raisons idéologiques, mais parce qu'elle seule est déterminée à remplir les conditions d'une énonciation collective. » (Kafka, 13, 15, 32, 33)

It appears that both Sartre and Deleuze feel that writing has the potential to liberate repressed populations, but their proper conceptions of what constitutes effective "political" aesthetic productions or creations diverge. We know that Sartre himself, an intellectual engagé, who participated in the French resistance, in student demonstrations such as that of May 1968, was convinced that large scale, overt movements of solidarity would effect social change. As for Deleuze, one can say he is no-less political but it would seem that as opposed to Sartre, Deleuze holds a stronger conviction in the strength and effectiveness of *molecular* political movements. The following section will suggest one of the influences for this conviction and will also clarify the sources of Bataille's and Deleuze's insistence on the pragmatic, empiricist associations with Wahl's "art of the *ET*."

JEAN WAHL'S AMERICAN THOUGHT AND DELEUZE'S PRAGMATIC PROJECT

> « La pragmatique (ou schizo-analyse) peut donc être représentée par les quatre composantes circulaires, mais qui bourgeonnent et font rhizome. »
>
> "The pragmatique (or schizo-analysis) can therefore be represented by the four circular components, but which bloom and make rhizome." (Mille Plateaux, 182)

Jean Wahl and Gilles Deleuze obviously shared an enthusiasm for Nietzsche's thought. Jean Wahl not only articulated his Nietzschean proclivities to Deleuze as professor, but eventually these two men stood together as colleagues to contribute to the Nietzsche Colloquium held at Royaumont in 1964. Despite the importance of this first link, one cannot overlook Deleuze's emphasis on the significance of Wahl's role in introducing "American thought" to French intellectuals. Deleuze's words furthermore imply that this American "thought" facilitated Wahl's capacity to generate and model the method of the "ET," for underlining the importance of the ET and its relation to the "dehors" to "rhizome": « *ET sous-entendu, rapport avec le Dehors, culte de la route qui ne s'enfonce jamais, qui n'a pas de fondations, qui file à la surface, rhizome.* » *ET double-entendre, relation to the Outside, cult of the road which never gives way, which has no foundations, which runs on the surface, rhizome.* (*Dialogues,* 73). As Bataille tells us, the pragmatic element in Wahl's thought lends it the unique, fluid, uncontainable quality that generates differences that allow Deleuze and others to think entirely new things in the French language. What aspects of American thought *think* differently from that of the French? When one peruses Wahl's writings on American philosophical approaches, one American figure and thinker stands out amongst all others. This figure is William James. Wahl writes several books that refer to American Pragmatism and Pluralism. In those books, such as *the Philosopher's Way* where Wahl outlines various figures who contributed to different philosophical movements, he mentions only William James in relation to Pluralism. Moreover, in Wahl's book, *Les Philosophes Pluralistes (1920),* only one of its four chapters is dedicated to a single thinker, being again, William James. In fact one can make the case that Wahl's Jamesean pluralism also counters and lends a way out of negativity, out of Hegelian dialectics. Pragmatic pluralism bears itself out of a disposition to see the world in its flux and diversity. The fluctuations and exchanges between heterogeneous elements result in an ever-evolving universe. In *Les Philosophes Pluralistes* (1920), Wahl expresses this: "When we observe the influences that contribute to pluralism, we see that most of these philosophers which inspire it, belong to a large scale reactionary movement against Hegel's doctrine. . . . These opponents of monists say that monists want to find peace for the soul in an abstract and general unity, the pluralist rather insists with love on distinctions and differences."(PP, 240). Many ideas in James came to fruition from countless influences, many of whom Deleuze studies and writes about such as Bergson, Emerson and Whitman. Due to this, the effort to draw a correlation between pinnacle ideas in Wahl's discussion of James's particular brand of pluralism, and similar themes in Deleuze's work is purely a speculative gesture. Regardless of the degree of accuracy in these speculations, it is certain that what Wahl did underline as interesting, unique and important in James's thought had to have only increased Deleuze's awareness, attraction to, and consideration of common themes, articulated by each one of these thinkers.

The ideas that Wahl underlines as important in James's thought that resonate with passages found in Deleuze's text involve terms and ideas such as, flux, fusion, multiplicities, the outside, the in between, limitations of intellectualism, the "cult of the individual," the limitless potential of the subject or variations of identity. Perhaps what becomes one of the most important elements Deleuze finds in this *introduction* to James via Wahl is the particular political sense that signifiers such as flux, mutations, destruction, creation, and movement, come to take on. In fact the case will be made that the kind of ideas generated by William James along with the creative process of the eternal return, enable Deleuze to *conceptualize* a different mode of being, which moves from *being* political to *becoming* political. These ideas that come out of this conceptualization also create the "Deleuzian" vision of a "political," writing and writer that counters those ideas forwarded by Sartre.

In *Vers le Concret,* ironically the same book that influenced Sartre while he was still writing as an individualistic young man and artist, contains a passage written by James that sounds alarmingly Deleuzian. In this section of the book, Wahl documents William James's response to the Dreyfus Affaire, which bears testimony to James's cosmopolitanism and to his idea of a universal democratic ideal. In response to the French "affaire" James theorizes on the role of the "intellectual" in political matters, James writes:

> "The work of intellectuals is always and everywhere to watch over to maintain the cult of the individual. From all of that little by little releases the idea that "large organizations" are dangerous. "I am against thickness and grandeur in all of their forms. I put myself on the side of invisible and molecular forces which operate from individual to individual, which file through cracks and crannies, as soft, tiny rivers or as the small, seeping capillaries of water; and all the same, if you give them time they destroy the most solid monuments of human arrogance."
>
> « Le devoir des intellectuels est partout et toujours de veiller au maintien du culte de l'individu. De tout cela se dégage peu à peu cette idée que les « grandes organisations » sont dangereuses. « Je suis contre la grosseur et la grandeur sous toutes leurs formes. Je me mets du côté des forces invisibles et moléculaires qui travaillent d'individu à individu, qui se faufilent à travers les fissures, comme autant de douces petites rivières ou comme les petits suintements capillaires de l'eau; et pourtant, si vous leur en laisser le temps, elles détruisent les monuments les plus solides, de l'orgueil humain. » (Vers le concret, 79).

One suggests that this Jamesean valorization of "molecular," "invisible" forces, along with other sources result in Deleuze's distinction between macro- and micro- political processes and creative activities. In his description of the micro-political, Deleuze uses similar metaphors and comes to the same result, being that the micro-political elements work as invisible forces, as molecular particles, as small lines, 'les fissures,' like small rivers or capillaries passing through the middle, imperceptibly but nonetheless wearing down the foundations of large organizations that threaten the cult, the freedoms, truths, realities, and variations of the individual. These micro-elements finally perhaps never destroy the macro-political structure that spawns oppressive organizations, and monuments, but they do cause binary categories that project static reflections of identities to slowly and subtly shift and lose credibility. « Bref, tout est politique, mais toute politique est à la fois macropolitique et micropolitique. » (In sum, all is politics, but all politics are both macropolitical and micropolitical.) Deleuze writes that fear often keeps the individual from countering the macro-system because it provides the security of constant and concrete ideals. Within this paragraph he describes its structure as one of large organizations, binary machines, and stability.

> "Fear, we can guess what it is. We are always afraid of losing. Our security, the great molar organization that sustains us, the arborescences we cling to, the binary machines that give us a well-defined status, the resonances we enter into, the system of over coding that dominates us- we desire all that. The values, morals, fatherlands, religions and private certitudes our vanity and self-complacency generously grant us. . . . We flee from flight, rigidify our segments, give ourselves over to binary logic . . . we reterritorialize on anything available. . . . The more rigid the segmentarity, the more reassuring it is for us." (Thousand Plateaus, 227)

> « La peur, nous pouvons deviner ce que c'est. Nous craignons tout le temps de perdre. La sécurité, la grande organisation molaire qui nous soutient, les arborescences ou nous nous accrochons, les machines binaires qui nous donnent un statut bien défini, les résonances ou nous entrons, le système de surcodage qui nous domine, nous désirons tout cela. Les valeurs, les morales, les patries, les religions et les certitudes privées que notre vanité et notre complaisance à nous-mêmes nous octroient généreusement. . . . Nous fuyons devant la fuite, nous durcissons nos segments, nous nous livrons à la logique binaire . . . nous nous reterritorialisons sur n'importe quoi. . . . Plus la segmentarité sera dure, plus elle nous rassure. » (Mille Plateaux, 277)

The struggle between stability and freedom takes place on all levels and in all areas: On the level of the individual, on the level of large "molar" organizations, in the area of art, philosophy, government, history, etc. While "les lignes de fuite," and molecular micro- currents move towards anarchy, the hard, molar macro-segments try to curtail their processes and "reterritorialize" them back into the sense and organization of the stabilizing system.

> "On the other hand, at the other pole, there is an abstract machine of mutation, which operates by decoding and deterritorialization. It is what draws the lines of flight: it steers the quantum flows, assures the connection-creation of flows, and emits new quanta. It itself is in a state of flight, and erects war machines on its lines. If it constitutes another pole, it is because molar or rigid segments always seal, plug, block the lines of flight, whereas this machine is always making them flow, "between" the rigid segments and in another, submolecular, direction. But between the two poles there is also a whole realm of properly molecular negotiation, translation, and transductions in which at times molar lines are already undermined by fissures and cracks." (Thousand Plateaus, 223)

> « D'autre part, à l'autre pole, il y a une machine abstraite de mutation, qui opère par décodage et déterritorialisation. C'est elle qui trace les lignes de fuite: elle pilote les flux à quanta, assure la création -connexion des flux, émet de nouveaux quanta. Elle est elle-même en état de fuite, et dresse des machines de guerre sur les lignes. Si elle constitue un autre pole, c'est parce que les segments durs ou molaires ne cessent pas de colmater, de boucher, de barrer les lignes de fuites, tandis qu'elle ne cesse de les faire couler, « entre » les segments durs et dans une autre direction, sub-moléculaire. Mais aussi entre les deux pôles il y a tout un domaine de négociation, de traduction, de transduction proprement moléculaire, ou les lignes molaires sont déjà travaillées par des fissures et des fêlures. » (Mille Plateaux, 273)

This Jamesean politics, that maintains the "cult of the individual," that finds its force through the energies conducted through individuals on a "microcosm level" would then condone Deleuze's assertion that the *individual* as "known" must "take flight" in order to move imperceptibly. As already mentioned, taking flight for Deleuze involves becoming other, others which stimulates a general becoming-other of society. To recall Sartre's idea of the writer's responsibility to be his/her-self, to write for one's group and with the *other's* perception of his/her social situation in mind, will again shed light on Deleuze's difference.

> "When the writer believes himself to have openings unto the eternal, he is mistaken . . . if he has understood that the surest way to be tricked by his time is to turn his back on it, or to pretend to rise above it. One does not transcend one's time in escaping it but in assuming it to change it."
>
> « Quand l'écrivain croit avoir des ouvertures sur l'éternel, il est hors pair . . . s'il a compris que le meilleur moyen d'être roulé par son époque c'est de lui tourner le dos ou de prétendre s'élever au-dessus d'elle et qu'on ne la transcende pas en fuyant mais en l'assumant pour la changer. » (Situations II, 257).

In accordance with the title of his book, *Situations II,* Sartre reveals again and again that "his" writer lives more bound than free. He expresses that the writer who chooses to write out of his status of freedom must bear the responsibilities that this "craft" implies. The writer must respond to certain expectations that society has of him. He must write of his problem, his time, for "his people," whether Jew, black, working class, etc. According to Sartre the writer can only change a social situation, avoid being "jipped" by his/her "époque" when he/she claims his/her historicity, assumes it rather than trying to transcend it or "escape" it.

Deleuze seems to respond directly to Sartre and refute these notions in both *Mille Plateaux* and *Dialogues.* Deleuze writes that the notion that equates taking flight with cowardice is based on "French ideas." Instead of a renunciation of action, taking flight rather activates the most effective of movements, because the energetic currents of lines of flight or lines of disappearance deterritorialize the foundations, the ideas, the misconceptions of oppressive power structures. For the flight itself makes other things take flight, for example, dominant systems and those elements that comprise them which lend them credibility and authority. D. H. Lawrence is cited to convey this idea. "The highest aim of literature, according to Lawrence, is 'To leave, leave, escape . . . to cross the horizon, enter into another life.'" The line of flight is a *deterritorialization.* "Anglo-American literature constantly shows these ruptures, these characters that create their line of flight, who create through a line of flight." Again, Deleuze writes that the French can't grasp this because they associate "fleeing" with desertion: With a desire to avoid commitments or responsibilities. Deleuze emphasizes that on the contrary nothing calls for more engagement and *self*lessness and gains the same kind of ground than tracing lines of flight: Only through flight does one discover and potentially create alternative worlds. (Dialogues, 47)

To resume, although both Sartre and Deleuze point to Kafka and American literature as providing models for a literature of the future their ideas on how the writer and the writing works to generate future truths, meanings and "revolutions" conflict. Deleuze's writer, contrary to what Sartre calls for, must forget, must enter a *becoming* non-writer, in order to be open to the plurality of lives that "give" and stimulate creative desire. In turn this "becoming-*other(s)*- writer" offers alternative lives and worlds that may be discerned through the proliferation of the writing's life-bearing particles, entities, lines, that can not be fixed in terms of Being. Deleuze juxtaposes two kinds of writing; one reterritorializes on fixed concepts by conforming to a code of dominant utterances, to a territory of established states of things, the other writing moves *with* and activates becomings. To achieve the latter mode of writing, it is imperative that the writer enter a becoming non-writer so that the creation becomes something other than writing. "Writing has no other goal: wind, even when we do not move . . . 'keys in the wind to set my spirit to flight and give my thought a gust of air from the backyard.' to release what can be saved from life . . . to release from the becoming that which will not permit itself to be fixed in a term." (Dialogues II, 74, 75)

In the above passage, Deleuze again speaks of a "courant d'arrière cour." This expression which he uses to describe Sartre as an influence along with the passage that describes Sartre as a milieu seems to contradict all that has been said on Sartre in the second half of this chapter. However, on the contrary, Sartre is *milieu* and *un courant d' arrière cour.* As Wahl tells us there are many Sartres in conflict and contradiction, there is the Sartre of *La Nausée,* of *Les Mouches,* of *L'Etre et le néant,* Sartre as literary artist, as literary critic and as philosopher. As Wahl stresses incessantly in his writings on Sartre's existential project and at the Colloquium at Royaumont when he speaks of Nietzsche; paradoxes and tensions result from attempts to articulate the inarticulable.[11] The seeming contradictions reveal the absent or silent irreducible truths that spring forth from the middle the in-between. Sartre similar to Nietzsche provides Deleuze with those "in between" passages or lines which inspire and offer "keys in the wind," those inexhaustible elements that resist being fixed within a term or concept. Deleuze writes through Sartre and Wahl *à l'Américain.* Instead of trying to interpret Sartre or Wahl in the traditional critical fashion, he extends their lives into the future where they increasingly 'become.' Their "lines of flight," extend up and over the void where they dropped off. This approach to his mentors resembles Deleuze's description of the workings of Anglo-American literature in *Dialogues* where Deleuze again contrasts English and Americans artists to French authors. Deleuze writes that as opposed to the French, who begin with a clean slate, tabula *rasa* and take up a primary certainty as a point of origin, the Anglo-Americans take up interrupted and broken lines and make them pass between points or obstacles. In this writing it is this in-between that is interesting not the beginning and end as they are points. In overthrowing the importance of "points" Anglo-American literature became a process of experimentation and "killed interpretation." (Dialogues II, 39, 49) What Deleuze finally produces out of his encounters with his "masters" and teachers, Sartre and Wahl is "Real philosophy"; a writing of love. As Gilles Deleuze states, all writing that creates in conjunction, which intersects with other elements in flux taken in a process that is simultaneously destruction and creation generates a "Difference," letters of love, conceptions of love. "Writing carries out the conjunction, the transmutation of fluxes, through which life escapes from the resentment of persons, societies and regimes. . . . One only writes through love, all writing is a letter of love: Real-literature." (Dialogues, 62)

Chapter Two

Disseminating the "Eaches": W.E.B. Du Bois's *Souls* and the Micro-Politics of Sound

> « Rhizomatique = Schizo-Analyse = Strato-Analyse = Pragmatique = Micro-Politique » (*Mille Plateaux,* 33)
>
> "Thus pragmatics (or schizoanalysis) can be represented by four circular components that bud and make rhizome. . . ."Behind" statements and semioticizations there are only machines, assemblages, and movements of deterritorialization that cut across the stratification of the various systems and elude both the coordinates of language and of existence. That is why pragmatics is not a complement to logic, syntax, or semantics; on the contrary, it is the fundamental element upon which all the rest depend." (Thousand, 146, 148)

As already noted, Deleuze writes that American literature exemplifies the meaning of *livre-rhizome:* the kind of book that disrupts the order of the *livre-racine* / book-root. Deleuze explicitly equates this movement with making a *pragmatics.* Pragmatics moves between things, with the logic of the ET, to upset ontology, by un-grounding its foundations and nullifying *major* points. As stated in the last chapter there is a reason for Deleuze's constant use of the term pragmatics. This reference to the pragmatic process as one that destabilizes the tree/root structures finds much inspiration in Wahl's text, *Les Philosophes Pluralistes,* which interprets and praises James's brand of pluralism or radical empiricism.

The first section of this chapter continues the thread of discussion that concerns Wahl's interpretation of William James's "new philosophy" and considers Deleuze's enthusiasm for and application of several of the ideas he finds there. Several questions will be discussed to indicate why these ideas become crucial components of Deleuze's thoughts on art and philosophy. These questions include; what aspects of James's "pragmatic pluralism" does Wahl underline as ground-breaking and radical? Where does one read these "radical," irreducible, pluralist "parts" at work in what Deleuze names his own "pragmatique" or "rhizomatique."

The remaining content of the chapter considers Deleuzian twists, terms and other "ET" elements that resist "logic" based rationalizations and defy the paradigms of totalizing schemas, in light of Wahl's transmission of *American thought,* to clarify their implications in application. One reads W.E.B. Du Bois's *The Souls of Black Folk* through certain Deleuzian concepts that evolve out of this intersection between James and Wahl among others, to hopefully activate certain "pragmatic" elements and techniques at work within Du Bois's text, elements that otherwise may go unnoticed, unheard or unappreciated.

Jean Wahl speaks of James's vision as one that favors movement and multiplicity over a monistic system such as that of Hegelian dialectics. "When we observe the influences that contribute to pluralism, we see that most of these philosophers which inspire it, belong to a large scale reactionary movement against Hegel's doctrine. . . . These opponents of monism say that monists want to find peace for the soul in an abstract and general unity, the pluralist rather insists with love on distinctions and differences."(PP, 240). Jean Wahl upholds pluralism as an innovative response to the synthesizing, reductive process of dialectics. This revolutionary philosophy includes the power of the sensorial, the strength of the will along with the intellect, to liberally "realize" and express an infinite number and range of experiences of the world. This irreducible approach also features *radical* mystical elements. James radical empiricism qualifies true thought as that which always mutates within a perpetual process of change, flux and flow that activates interconnections and exchanges between infinite, heterogeneous elements transmitted through different consciousnesses. Rather than feigning to solve the mysteries of existence(s), lives, consciousness' in the universe, pluralism understands and appreciates the complex nature of all life forms and "worlds" and consequently advocates that "problems" or mysteries should be articulated and entertained, but often should remain unanswered or unsolved. According to James and Wahl, more often than not the "answers" only reduce what can't be known, and freeze frame or deny the reality of invisible radical forces and movements in the world to preserve the integrity of ontology based systems.

> "What attracts in pluralism is this vision of a multiple and moving world made up of irreducible wills in combat. . . . First off if appears in pragmatic works of philosophy that there is a new conception of philosophy. Philosophy not only thinks with thought, but with its feeling and with its will to discover new horizons. In a very original way, James united an original theory of the will, an irreducible empiricism and mysticism. . . . Pluralism no longer appears to us as a solution to metaphysical problems; it poses problems; and it wants certain problems to remain posed, instead of solved, or in other words, to be finally annihilated in the absolute. The principle criticism that it addresses to monism is that monism destroys certain ideas in transforming them. . . . Pluralism is rather the affirmation of the irreducibility of certain ideas and things." (PP, 242, 243)

> « Ce qui attire dans le pluralisme, c'est cette vision d'un monde multiple et mouvant fait de volontés en lutte, irréductibles les unes aux autres. . . . Il semble d'abord qu'il y ait dans les oeuvres des philosophies pragmatistes une conception nouvelle de la philosophie. La philosophie ne pense pas seulement avec sa pensée, mais avec son sentiment et avec sa volonté. Et il découvre des horizons nouveaux. James unissait d'une façon originale une théorie de la volonté, un empirisme irréductible et un mysticisme. . . . Le pluralisme ne nous apparaît pas comme une solution aux problèmes de la métaphysique ; il pose des problèmes ; et il veut que certains problèmes restent posés, qu'ils ne soient pas résolus, c'est à dire finalement annihilés dans l'absolu. Le principal reproche qu'il adresse au monisme, c'est que le monisme en les transformant détruit certaines idées. . . . Le Pluralisme est donc l'affirmation de l'irréductibilité de certaines idées et de certaines choses. » (Philosophes Pluralistes, 242, 243)

This passage echoes many ideas articulated in Deleuze's texts. Firstly it speaks to Deleuze's perspectives that appreciate and view the "world" as "becoming," multiple, mutating, and moving, through a struggle and combination of diverse forces. Secondly, in *Mille Plateaux* Deleuze names the ultimate philosophical procedure and deterritorializing plan a "pragmatique," and characterizes its operation as one that makes things bloom and "makes rhizome." Deleuze like Wahl also stresses that in a deterritorialized *philosophy* it is more important to present problems rather than pretending to solve them. « La vérité est qu'il s'agit, en philosophie et même ailleurs, de trouver le problème et par conséquent de le poser, plus encore de le résoudre. » (*The truth is that in philosophy and in other domains, it is about finding the problem and therefore posing it, much more than it is about solving it.*) (Magazine *Littéraire,* 31)

According to Wahl, the *relations* in James's empiricism are what qualify it as "radical." « Mais James ajoute que c'est un empirisme radical parce qu'il prête attention, non pas seulement aux faits, mais aussi aux relations entre les faits. » (*James adds that it is a radical empiricism because it pays attention not only to acts/facts but to the relations between them.*) (PP, 122). The "relations" or what Wahl also refers to as the "avec" in quotes, the conjunctive forces, are radical because they are so essentially diverse and independent that "one can imagine a world of only the "avec" and nothing but." (PP, 123) Wahl's description of the "avec," elements which move every which way "exterior to terms," and relate heterogeneous elements while they themselves hold multiple relations within themselves, definitely exemplifies the "ungraspable" quality of Wahl's work as Georges Bataille characterizes it. However, it is precisely because the operations of the "avec" exceed logical systems, and generate non-logical worlds, that they cannot be rationally explained or defined.

Wahl emphasizes that these relations affirm and reflect the chaotic state of the pluralistic universe. In this world the relations merge, connect, unravel, knot together, and come undone. These relations move from simple simultaneity, from simple "avec," as far as resemblance, to inter-activity, finally to relations between states of consciousness, even to the absolute continuity of streams of consciousness. (PP, 123–124) Wahl explains that due to these "relations" it becomes possible to understand how one's consciousness inserts itself into the "exterior world" in a-temporal moments. For this *exterior world* offers relations with all elements closed off by barriers such as spatial boundaries or "chronological" notions of time. Within this vision *Being* itself always moves, becomes and mutates since "no being contains all the others." Being therefore conjugates with an infinite array of "outside" "avecs" that always escape the system. In this sense Being mutates through the transformational process of the avec, mirrored in Deleuze's description of the operations of the "*Et . . . Et . . . Et.*"

> "There is therefore a multitude of relations of 'conterminosity,' of confluence. When we see minds that know a like thing between them, we are in the presence of this experience of confluence. From this, one can understand how our consciousness inserts itself in the exterior world from time to time, in discontinuous moments at that. All sorts of floating, varying,

> and free relations can take place between things- continuity, resemblance, simultaneity, proximity, superposition. . . . 'Pluralism pragmatically considered is, James explains, the idea that different parts of reality can have relations purely exterior to themselves, between themselves.' And finally, as vast as the being considered, there is always an "outside." There is not a being that contains all the others, there is always something that escapes, which does not want to come back into the system."

> « Il y a ainsi dans le monde une multitude de relations de « conterminosité, » de confluence. Quand nous voyons des esprits qui connaissent une même chose entre elles, nous sommes en présence d'expériences de confluence. Dès lors, on peut comprendre comment notre conscience s'insère dans le monde extérieur de temps à autre, à des moments discontinus aussi. Il peut y avoir toutes sortes de relations flottantes, variées, libres, entre les choses—continuité, ressemblance simultanéité, proximité, superposition. . . . « Le pluralisme considéré pragmatiquement, est, dit William James, l'idée que les différentes parties de la réalité peuvent avoir entre elles des relations purement extérieures. » Et enfin, si vaste que soit l'être considéré, il y a toujours un « en dehors. » Il n'y a pas d'être qui contienne tous les autres, il y a toujours quelque chose qui échappe, qui ne veut pas rentrer dans le système. » (Philosophes Pluralistes, 124, 5, 7)

Deleuze instills an element of mysticism that resembles that which Wahl's text presents. Deleuze insists that becoming *happens* through a process that incites intersections with other diverse and ever-changing elements. This process may also be viewed as one stimulated by spiritual forces, the will, the mind, or a combination of all three and as one that relies on a world of movement, flux, "multiplicities," variant lines, particles, effects and the "avec" to create the fertile soil of the *in between,* the milieu. As Deleuze states in *Dialogues,* the writer creates a world but not one that is already there, that waits to be created. This happens as the author "writes with," "speaks with"; *with* the world, with "people." This communication is not a conversation but a conspiracy and collaboration. "*All the subtle sympathies of the soul without number.* That is agencing: being in the middle, on the line of encounters of both an exterior and interior world." (Dialogues II, 52)

This citation and others in *Dialogues* and *Mille Plateaux,* speak of the artist or the "artist-philosopher" as he/she who creates new worlds through the process Wahl describes as "conterminosité." The each one, the particular in an "everyone" grouping which a "molar ensemble" conceptualizes, breaks free and releases its particles and affects. These liberated forces displace centers of gravity, and eliminate all that "enracinates" the *each one* in a static, "everyone" block, through abstract interconnecting lines that perpetuate further "conjugations": This results in a different world, where the released "each ones" participate in a becoming that destroys the idea of "everyone," as "block," homogeneous grouping.

> "Eliminate . . . all that roots each of us (everybody) in ourselves, in our molarity. For everybody/everything is the molar aggregate, but becoming everybody/everything is another affair, one that brings into play the cosmos with its molecular components. Becoming everybody/everything (tout le monde) is to world (faire du monde), to make a world (faire un monde). . . . The Cosmos as an abstract machine . . . enter the haecceity and impersonality of the creator. One is then like the grass: one has made the world, everybody/everything, into a becoming, because one has made a necessarily communicating world, because one has suppressed in oneself everything that prevents us from slipping between things and springing up in the midst of things."

> « Eliminer tout ce qui enracine chacun (tout le monde) en lui-même, dans sa molarité. Car tout le monde est l'ensemble molaire, mais *devenir tout le monde* est une autre affaire, qui met en jeu le cosmos avec ses composantes moléculaires. Devenir tout le monde, c'est faire monde, faire un monde. . . . Le Cosmos comme machine abstraite . . . entrer ainsi dans l'heccéité comme dans l'impersonnalité du créateur. Alors on est comme l'herbe : on a fait du monde, de tout le monde un devenir, parce qu'on a fait le monde nécessairement communicant, parce qu'on a supprimé de soi tout ce qui nous empêchait de nous glisser entre les choses, de pousser au milieu des choses. » (Mille Plateaux, 343, 344).

All this activity that takes place in this abstract "cosmos" machine, disperses innumerable *ones,* ones ever becoming, this destroys the molar world picture, a schema that forces and reduces all *radical* "outside" particles and effects into an equation where the "everyone"= one.[1] This passage resonates with Wahl's insistence that in James's pluralist philosophy the parts create the whole, whereas most philosophies define the parts in terms of the whole. In James's radical universe the whole takes shape only in relation to the parts. The parts' meanings moreover derive only from themselves.

> "This is an empiricist philosophy according to James, because the parts are more important than the whole, because the whole is not the sum of its parts, this is a philosophy of mosaics, a philosophy of plural facts/events, facts without support or substance other than themselves; whereas rationalism tends to explain the parts in terms of the whole, empiricism tends to explain the all by the parts, it sees the universe "distributively" and not collectively." (Philosophes Pluralistes, 122)

The elusive, irreducible elements, the *avec,* those *relations* in-between that Wahl takes such great pain to describe in the preceding passages, again operate identically to Deleuze's *ET* . . . *ET* . . . *ET* . . . that Deleuze grants such importance to: A passage in Deleuze's *Dialogues* manifests this as it describes the intricate nature of their "fluid and ungraspable" quality. [2] In this passage Deleuze equates the "exteriority of relations" with "AND" and emphasizes that the exteriority of relations is a *vital protest against principles* in that they activate an experimentation wherein the relations create a very strange world piece by piece constituted by blocs and ruptures, attractions and divisions, conjunctions and interweavings. To "experiment" in this manner one must be prepared to substitute the AND for IS (ET for EST). In other words one must suppress a reliance on ontological concepts and assumptions and be willing, alive and courageous enough to create in an unstable territory with whatever non-ordered elements one encounters. The AND usurps the authority of the "est" (to be), equated with philosophical "certainties" and the "molar order"; while it "worlds"; "makes relations shoot outside their terms and outside the set of their terms, and outside everything which could be determined as Being, One, or Whole." Not surprisingly Deleuze's adulation naming Jean Wahl "France's most important philosopher" concludes acclaiming the force and creative function of AND. Deleuze continues to praise and express gratitude for Wahl: "Not only did he introduce us to English and American thought, but had the ability to make us think, in French, things which were very new; he on his own account took this art of the AND . . . this minoritarian use of language, the furthest."(Dialogues II, 58).

The next section of this chapter reads W. E. B. Du Bois as "experimenter." As the kind of American, radical empiricist whose pragmatic enterprise does "violence to thought." Piece by piece, Du Bois creates a strange world full of conjunctions and disjunctions of a-logical substances. Du Bois weaves dissimilar elements together with the method of the ET which undermines the notion of one unified, fixed, colored identity, fabricated under the corresponding notion of a unified, homogeneous, black "everyone"-grouping. Du Bois's abstract lines like Deleuze's "lignes de fuite," escape the "Molar ensemble" that tries to define and englobe black "Being." Du Bois activates these lines through a style that releases a plurality of cacophonous, yet harmonious, disembodied voices releasing the non-material "essences" of "de-*race*-inated"/de-*racine*-ated" souls.

LIVRE-RACINE VERSES LIVRE-RHIZOME

> "There is a universe of truths which is one of the possible and the past; there one encounters dead truths and truths "before the word," there one finds melodies of future and geometric relations which are yet to be known."
>
> « Il y a une univers de vérités qui est l'univers du possible et du passé; là se rencontrent les vérités mortes et les vérités "avant la lettre," là on trouve les mélodies de l'avenir et les relations géométriques qui ne se sont pas encore connus. » *William James* (Philosophes Pluralistes, 140)

This writing now proceeds to consider Du Bois's, *The Souls of Black Folk* in relation to the anti-principle, principles of the pragmatics of the ET, to invite its readers to *think with "ET," instead of thinking "EST," to think for "EST" to chase away the terms and the ensembles, the One "groupings."* One first considers how Du Bois's "pragmatic" approach to writing, to expressing experiences in the world, indeed moves the *book-root* towards a becoming *book-rhizome.* In other words one will ask, how does *Souls* stylistically operate to actualize future possibilities out of a fusion of past materials and truths "before the letter" as James's describes? Truths not yet reduced by language, to result in the creation of yet unheard, future "melodies," and yet to be discovered "geometrical relations." In addition, going back to the ideas discussed at the end of the last chapter that distinguish between micro- and macro- political activities, one will ask questions concerning the political implications of Du Bois's work.

Deleuze explains in both *Mille Plateaux* and in *Kafka pour une littérature mineure* that writing in a "minor" way most forcefully advances a politically charged "minoritaire" movement. Minor literature disrupts the "Major order" as it deterritorializes its major language, and sounds "unmasterly," unidentifiable, narrative voices that invite the reader to participate in a politically charged, revolutionary, collective rebellion that activates its "weapons" through micro- channels, on micro-levels. Minor writing, the lines it extends destabilize the codes, structures, ideals and truths of the Major, molar system, which bases its perceptions of the world on what Deleuze calls the tree model.

"A minor literature is not one of a minor language, but rather one that a minority makes in a major language. The first characteristic is that the language is affected by a strong coefficient of deterritorialization. . . . The second character of minor literature it that all is political. . . . What the writer says alone already constitutes a communal action, and that which he says or does is necessarily political, even if others don't agree. . . . It is literature that produces an active solidarity, despite skepticism; and if the writer is in the margins or a side from his fragile community, than he is even more in the situation to express another potential community, to strike out the means of an other consciousness and an other sensitivity/sensibility."

« Une littérature mineure n'est pas celle d'une langue mineure, plutôt celle qu'une minorité fait dans une langue majeure. Mais le premier caractère est de toute façon que la langue y est affectée d'un fort coefficient de déterritorialisation. . . . . Le second caractère des littératures mineures, c'est que tout y est politique. . . . Ce que l'écrivain tout seul dit constitue déjà une action commune, et ce qu'il dit ou fait est nécessairement politique, même si les autres ne sont pas d'accord. . . . C'est la littérature qui produit une solidarité active, malgré le scepticisme; et si l'écrivain est en marge ou à l'écart de sa communauté fragile, cette situation le met d'autant plus en mesure d'exprimer une autre communauté potentielle, de forger les moyens d'une autre conscience et d'une autre sensibilité. » (Kafka, 30–33)

One can read Deleuze's description of "minor literature" as an extension of James's idea that concerns "molecular" political phenomena. Deleuze's minor-literature activates the Jamesean micro-political forces that are channeled from individual to individual to result in a psychically collective micro-movement that has the potential power to destroy the most "solid monuments" and the most ominous organizations. In the introduction of *Mille Plateaux,* Deleuze juxtaposes minor literature or livre-rhizome to livre-racine or major literature that instead reflects and reinforces the ideology behind macro institutions and thought systems.

"A first type of book is the root-book. The tree is already the image of the world, or the root the image of the world-tree. . . . The book imitates the world, as art imitates nature. . . . The law of the book is the law of reflection. . . . We find ourselves before the most classical and well reflected, oldest, and weariest kind of thought. Nature doesn't work that way: in nature, roots are taproots with a more multiple, lateral, and circular system of ramification, rather than a dichotomous one. . . . The world has become chaos, but the book remains the image of the world. . . . A strange mystification: a book so much more total than fragmented. The book as the image of the world, at any rate, what a vapid idea." (Thousand, 5, 6)

« Un premier type de livre, c'est le livre racine. L'arbre est déjà l'image du monde, ou bien la racine est l'image de l'arbre-monde. . . . Le livre imite le monde, comme l'art la nature. . . . La loi du livre, c'est celle de la réflexion . . . nous nous trouvons devant la pensée la plus classique et la plus réfléchie, la plus vieille, la plus fatiguée. La nature n'agit pas ainsi: les racines elles-mêmes sont pivotantes, à ramification plus nombreuse, latérale et circulaire, non pas dichotomique. . . . Le monde est devenu chaos, mais le livre reste image du monde, chaosmos-radicelle, au lieu de cosmos-racine. Etrange mystification, celle du livre d'autant plus total que fragmenté. Le livre comme image du monde, de toute façon qu'elle idée fade.» (Mille Plateaux, 11, 12, 13).

The livre-racine (book-root), reflects and reinforces the Molar Unity or the Tree-Structure on which it depends. The livre-racine models a thinking based on pre-existing forms, formulas, binary oppositions, logically, genealogically traceable, stable identities, truths and recordable histories. Terms associated with the livre-racine include; roots, rooting, enracination, a *logically* traceable, genealogy. The "tree model" standardizes a conception of Being as fixed in one unified identity that moves in a chronological order from past to present to future to *death,* A Being always in one direction. The book root follows the model of arborescent systems that define, structure, stabilize identity and reduce worlds to One.

"It is odd how the tree has dominated Western reality and all of Western thought, from botany to biology and anatomy, but also of gnosiology, theology, ontology, all of philosophy . . . the root-foundation, *Grund, roots et foundations.* . . . Aborescent systems are hierarchical systems with centers of significance and subjectification, central automata like organized memories. . . . The channels of transmission are preestablished: the arborescent system preexists the individual, who is integrated into it at an allotted place."(Thousand, 18, 16)

« C'est curieux, comme l'arbre a dominé la réalité occidentale et toute la pensée occidentale . . . toute la philosophie; le fondement-racine, *Grund,*

> *roots et fundations.* . . . Les systèmes arborescents sont des systèmes hiérarchiques qui comportent des centres de signifiance et de subjectivation, des automates centraux comme des mémoires organisées. . . . Les canaux de transmission sont préétablis: l'arborescence préexiste à l'individu qui s'intègre à une place précise. » (Mille Plateaux, 27, 25).

The tree model dominates the perceptions, ideas, and ideals of the west and its thought. It thinks in terms of roots and foundations, hierarchical systems which pre-establish all positions and relations. The classic book or book-root, refers back to and reflects life projected by the arborescent political, philosophical and historical systems. Again, contrary to Sartre's position articulated on page eighty-five of this text, Deleuze affirms that the problem is not how to exert one's freedom in the root-system but how to find a way out, an exit: In his book dedicated to Kafka's *littérature mineure or livre-rhizome* he writes, "As Kafka states, the problem is not about freedom, but about a way out . . . how to find a pathway there where one has not been found."(Kafka, 19). Finding an exit and tracing it out entails activating what Deleuze calls "les lignes de fuites." These lines stimulate intensities that generate other paths that move through and beyond the "whole structure" to result in "fuites"/leaks in the system. If one compares features of the book-root as opposed to what Deleuze calls the livre-rhizome that functions in the mode of the system-*radicelle,* then one can appreciate how the former book reinforces the structure and operations of the verb *être*/est where designated positions are fixed and defined to reflect the pre-existing structures that substantiate and arrange the "real" world order. However the latter model, that of the book-rhizome, operates as a radical empiricism that undermines notions based on pre-existing formulas. The livre-rhizome finds its ways out as Kafka does, by activating the invisible, molecular irreducible aspects of life and reality that gnaw at the foundations of unifying systems, and destabilize them.

> "The radicell-system, or fascicular root, is the second figure of the book. . . . The abortionists of unity are indeed angel makers, doctores angelici, because they affirm a properly angelic and superior unity. . . . Such a system could be called rhizome. . . . Principles of connection and heterogeneity: any point of a rhizome can be connected to anything other and must be. This is very different from the tree or root, which still plots a point, fixes an order. . . . A rhizome has no beginning or end; it is always in the middle, between things, interbeing, *intermezzo.* . . . The tree imposes the verb "to be," but the fabric of the rhizome is the conjunction,

> "and . . . and . . . and . . ." This conjunction carries enough force to shake and uproot the verb "to be." (Thousand, 5, 6, 25)
>
> « Le système -radicelle, ou racine fasciculée, est la seconde figure du livre. . . . Les avorteurs de l'unité sont bien ici des faiseurs d'anges, docteurs angelici, puisqu'ils affirment une unité proprement angélique et supérieure. . . . Un tel système pourrait être nommé rhizome. . . . Principes de connexion et d'hétérogénéité : n'importe quel point d'un rhizome peut être connecté avec n'importe quel autre, et doit l'être. C'est très différent de l'arbre ou de la racine qui fixent un point un ordre. Un rhizome ne commence et n'abouti pas, il est toujours au milieu, entre les choses, inter-être, intermezzo. L'arbre impose le verbe « être, » mais le rhizome a pour tissu la conjonction « et. . . . et. . . . .et . . . » Il y a dans cette conjonction assez de force pour secouer et déraciner le verbe être. » (Mille Plateaux, 12, 13, 36).

REPLACING UNITY WITH RHIZOME: DU BOIS *UN DES DOCTEURS ANGELICI*

W.E.B. Du Bois creates as one of Deleuze's *docteurs angelici* in the construction of his work, *The Souls of Black Folk.* In trying to follow the paths and movements of this rhizome system, *système-radicelle,* or radical cells, this reading tries not to impose normative standards, or sum up Du Bois's project in terms of any Whole(s): In terms of trying to establish some thematic unity or "goal" in relation to any "whole-structure," whether a whole historical époque, literary period, social grouping or from the artist's "whole" life or "whole life project." Despite this direction, one does not discount the importance of appreciating Du Bois's marginalized position that makes Du Bois a *problem.* In *Color and Culture* Ross Posnock states that Du Bois, *being a problem* results from living as oxymoron, as unclassified residuum in a world that polarizes the terms, intellectual and black. "White supremacy's identity logic was underwritten, said Du Bois, by "Nature's law" which decreed that "the word 'Negro' connotes 'inferiority' and 'stupidity' lightened only by unreasoning gayety and humor" . . . hence black intellectual became a repugnant oxymoron, a corrupt and decadent monstrosity."(Color, 58). Du Bois articulates the significance of his "problematic" position in the first paragraph of his opening chapter. "Between me and the other world there is ever an unasked question. . . . . How does it feel to be a problem? . . . being a problem is a strange experience,—peculiar even for one who has never been anything else, save perhaps in babyhood and in Europe."(Souls, 3, 4). Du Bois's words clarify that this "position" stems exclusively from the immediate "outside" white American world that surrounds him rather than from any internal or universal precondition. Du Bois poses a problem to white America, that founds itself on a hierarchical racial structure that polarizes the terms, black/white, inferior/ superior, primitive/civilized, intellectually inept/ and mentally adept. In this passage in *Souls* and many others in Du Bois's writings, he expresses that being a problem is exclusive to American terrain. In other nations such as France and Germany he is seen as a scholar, a man of letters, an intellectual who happens to be a Negro- American, and it is not a problem, he is not a problem.[3] Surely, in Du Bois's mind looms the question, can the American "world-order," its perceptions, laws, prejudices, murky distorted vision which judges on color rather than character be corrected? Through art can one establish an alternative, "*superior, angelic unity*" which *aborts* the underpinnings of the rooting orders, a prerequisite to live as non-problem, as an *each one* rather than as a *sub-one* or a Negro-*one.* This writing will show that *The Souls of Black Folk* articulates such questions and tries to force thought to think what is "*repugnant to thought,*" in attempt to alter, adjust, correct, the senses, and perceptions of the American major order. *Souls,* a text that includes verse and song demands an auditory appreciation. In accord with Deleuze's description of rhizomatic texts as ones that hold an inherent "musical" dimension, this reader follows his advice and listens to Du Bois's text as one listens to and "receives a song." [4] (Dialogues II, 3, 4). While it takes the sounds of the text, and other sensorial elements, into account, this reading tries to indicate to what extent *Souls* extols the principles of connection and heterogeneity to disrupt the notion of "Unity" and shake up and *de-race-inate* the verb "être." It especially hopes to convey the extent to which *Souls* develops the middle ground wherein new meanings bourgeon to resist an "End."

W.E.B. Du Bois who writes in a *pragmatic* fashion emphasizes in many places the influence of William James and the degree to which he revered him. In *Dusk of Dawn* he writes, "Of course in general philosophy under the serious and entirely lovable president was different. It opened vistas. It made me determined to go further in this probing for truth. Eventually it landed me squarely in the arms of William James of Harvard, for which God be praised. . . . I was repeatedly a guest in the house of William James; he was my friend and guide to clear thinking."(*Writings,* 578, 581).[5]

Du Bois describes the frustration he feels as an intellectual, who has earned a Ph.D. at Harvard, proven himself intellectually superior rather than equal to the average white person, yet nonetheless lives "shut out from *their* world by a vast veil." "Why did God make me an outcast and a stranger in mine own house?"(Souls, 5). Du Bois surely finds James's non-hierarchical, democratic, intellectual process that allows for multiple truths and realities in the making worthy of passionate fascination. This pragmatic philosophy that appreciates a variety of interpretations and expressions of the world, and even conceives of the possibility of multiple worlds coexisting or even colliding to evolve into new, more inclusive worlds promotes a realm where the minority can at least think, and contribute as an equal participant. Wahl underlines the importance of the "democratic" aspect in James's pragmatism that undoubtedly appealed to Du Bois whose concrete everyday "political" world experiences, dictated and informed by Jim Crow laws and other societal mechanisms, separate him from a larger world.

> "For James, all theories of the outside world that annihilate the idea of exteriority, will be necessarily inadequate . . . and in this world, there is not an immutable hierarchy; each being is of equal importance; each thing is on the same level. . . . The philosophy of the particular is a democratic philosophy. . . . The philosopher can even dream, aside, *beyond and above present realities,* an other world, other worlds. . . . Thanks to this vision of things, each moment will appear to us as a new universe."

> « Pour James, toute théorie du monde extérieur qui annihile l'idée d'extériorité, sera nécessairement inadéquate. . . . Et dans ce monde, il n'y a pas de hiérarchie immuable; chaque être a une égale importance; chaque chose est sur le même rang. . . . La philosophie du particulier est une philosophie démocratique. . . . De même le philosophe pourra rêver, à côté, *au-dessus des réalités présentes,* un autre monde, d'autres mondes. . . . Grâce à cette vision des choses, chaque moment nous apparaîtra comme un univers nouveau. » (Philosophes Pluralistes, 106, 107).

James's pluralism affirms what Du Bois experiences as a painful reality. Different worlds do in fact co-exist. Du Bois's world is one that the white world pushes back and closes off from the democratic promises of equal opportunity, status and freedom. However, James's democratic philosophy also affirms the legitimacy of Du Bois's separate but not less-than sense of reality and empowers and encourages him towards action. As already described, James's pluralism begs for an upheaval of the orders that fix individuals in limiting, narrow worlds and provides a method to be activated. In a Jamesean "pluralistic universe," power comes from the minute particulars, the sometimes silent and invisible forces, or those sounds and sights that society stifles and effaces from view. In this universe the individual who can see beyond certain absurdities instituted by the "Molar World," can undermine their legitimacy through discursive and subversive expressions of the will, and the senses. This person must stimulate and renew muddled sensual perceptions dulled by the imposition of standard filters congruous with the *Major Order.* Possibilities for change are limitless as one activates subtle vehicles that travel through the fissures and minute capillary like channels. A staid sense of reality, Being, or experience, can be overthrown as micromovements create collective mutations wherein beings and worlds shift towards becoming beings and worlds in the making.[6]

In order for Du Bois's creative project to be effective it necessarily needs to move beyond a purely intellectual, rational sphere and its linguistic register. James states that to describe, the life and realities that exceed language and its concepts that cut, separate and fix terms, and consequently negate meanings that operate "illogically," that move *between* elements that theoretically don't connect, requires some medium of expression other than 'talking.' Similar to Deleuze, James tells his students at Harvard, to listen to his lecture passively, as one "receives a song." In order to "feel," sense the "realities" and meanings of a *Pluralistic Universe,* one must listen receptively and resist reactions and judgments fueled by preconceived formulas and conceptualizations. James tells his students that they must shut out external and internal "verbal clatter" to think in non-conceptual terms which may involve an "inner catastrophe."(Pluralistic, 290). This experience demands a particular disposition, one that can accept the "unknown." "Not everyone is capable of such a logical revolution."(Pluralistic, 266). "I know that by vainly seeking to describe by concepts and words what I say at the same time exceeds either conceptualization or verbalization. As long as one continues 'talking,' intellectualism remains in undisturbed possession of the field. The return to life can't come about by talking. It is an act; to make you return to life I must deafen you to talk. . . . The minds of some of you, I know, will absolutely refuse to do so, refuse to think in non-conceptualized terms." (Pluralistic, 290) In order to present his "other" world, Du Bois's writing must operate contrary to the forum of a traditional book that reflects the *Major Order,* of the majority "white world," whose ideologies and concepts render him, and those he represents, inferior, sub-citizens. Du Bois's book-rhizome writes differently than the book-root to both resist being reduced to a narrow spectrum of meanings and in order to undermine the foundations of that "*majoritarian*" world, and reveal the injustice inherent in its legalized–criminal, social abuses and mechanisms. Du Bois, like James in his lecture, urges that one listens and *hears* his book, rather than reads or interprets it with the values of the major order. The reader must willingly receive *half-articulate* yet profound messages in a realm beyond the major orders' codes and definitions, in that world on the "black" side, the other side of what Du Bois names the "Veil"; that pluralistic metaphorical apparatus that among other things signifies the divider, a segregation mechanism that separates individuals via the color line.

Du Bois's unique composition that comprises, lyrics and musical notations, lines from black spirituals, along with, snippets of poems, historical accounts, sociological synopses, autobiographical sketches and a short story, creates an organic (rhizomatic) rather than logic oriented text. Donald Gibson points this out in the introduction to the Penguin classics edition. "The relationship among the essays is largely organic, rather than logical. . . . Du Bois's work does not lend itself entirely to a logical analysis. . . . This is because of Du Bois's awareness of the strengths and limitations of logic. . . . He felt the need, to seek methods of persuasion beyond the logical. These he found in metaphor." (Souls, viii, ix). One may say in conjunction and continuation that Du Bois recognizes the signifying power of a text that comprises a panoply of various genres.[7] Such a text generates meanings that prove ever *organic* in that it resists and exceeds normative, reductive interpretations. Du Bois non-standard mélange produces the kind of work Jacques Derrida describes in "Plato's Pharmacy." "A text is not a text unless it hides from the first comer, from the first glance, the law of its composition and the rules of its game. . . . There is always a surprise in store for the anatomy or physiology of any criticism that might think it had mastered the game surveyed all the threads at once."(Pharmacy 63). Du Bois communicates throughout his text that its meanings are not to be had at first glance. Du Bois first announces this in his *Forethought* saying, "many things," "strange meanings," "grains of truth," "deep recesses," lie buried, hidden and must be sought, excavated, to be "dimly discerned," and all of them signify and operate on their own terrain, that of the *other* side, the *black* side of the Veil. "HEREIN LIE buried many things . . . the strange meaning of being black here in the dawning of the Twentieth Century. . . . Leaving then the world of the white man, I have stepped within the Veil, raising it that you may view faintly its deeper recesses." (Souls, 2). Upon exiting the white world Du Bois's narrator lifts up the Veil and invites his readers to join him to move into that other world that composes the book and gives it its meanings. Only by accepting Du Bois's invitation that surely entails the sort of mind-blowing, *inner catastrophe* James speaks of, can one appreciate the depth, the language, the sounds and significance of Du Bois's book.[8]

THE MICRO-POLITICS OF THE *SORROW SONGS*

> "The crack happens almost without one knowing it but one becomes conscious of it suddenly indeed." This molecular line, more supple, but no less disquieting, in fact, much more disquieting, is not simply internal or personal: it also puts everything into play, but on an other scale and in different forms, with segmentations of a different nature, rhizomatic instead of arborescent. A micro-politics." (Thousand, 199)

> « La fêlure se produit sans presque qu'on le sache, mais on prend conscience vraiment d'un seul coup. » Cette ligne moléculaire plus souple, pas moins inquiétante, n'est pas simplement intérieure ou personnelle : elle aussi met toute chose en jeu, mais à une autre échelle et sous d'autres formes, avec des segmentations d'une autre nature, rhizomatiques au lieu d'arborescentes. Une micro-politique. » (Mille Plateaux, 243)

This chapter section will illustrate how Du Bois's text proliferates lines and sounds that put a variety of elements into play but within a different order or system, that generates and values non-conceptual meanings and forms. It will also show the extent to which this type of "micro-political" creation is susceptible to distortions and mis-readings due to violating interpretive practices that "reterritorialize." In other words, "rhizome books" with their discursive, elusive, molecular movements and strange *other* world meanings, may find themselves easy prey to interpretations that try to coerce them into pre-existing, "reasonable" forms and concepts generated by the structures they strive to break both from and down. Reterritorialization therefore snuffs out a-logical signs and nullifies their non-conceptual "sense." This chapter both traces out deterritorializing, rhizomatic movements, and reveals certain approaches that negate these movements to shut them or their meanings down. Ultimately one hopes to underline the value of *pragmatic* processes of reading when dealing with micro-political productions, particles and particulars.

Donald Gibson names the Veil the central, unifying metaphor in Du Bois's text. However this text disagrees and forwards that the "CRY" that sounds itself from beyond the Veil resonates as the most profound and insistent element that links the book's chapters. Du Bois urges that if one hears, listens to the "cry" than the book's meanings will potentially set the "imprisoned free." "HEAR MY CRY, O God the Reader, vouchsafe that this my book fall not still-born into the world-wilderness. . . . If somewhere in this whirl and chaos of things there dwells Eternal Good, pitiful yet masterful, then anon in His good time America shall rend the Veil and the prisoned shall go free . . . free as yonder fresh young voices welling up to me from the caverns of brick and mortar below–." (Souls, 215, 217). Only if the reader accepts to enter the *other* side, the *black* side of the Veil will he/she "HEAR" Du Bois's "CRY" and appreciate its potency and organic vitality. The "Cry" sounds itself from within the *music* of the text, especially within that music which effuses forth from the *Sorrow Songs.* Du Bois emphasizes the importance of the songs throughout the text and finally immerses the reader within them in the concluding chapter dedicated to and entitled, "The Sorrow Songs." The songs stand not only as "America's only music," but represent the nation's only *spiritual heritage* suggesting that the "soul," center, and foundations of *white* America are inherently spiritually bankrupt. The black spiritual translates a redemptory path through which to readjust the American soul, to attain salvation. In addition, these songs represent the most profound expression of human experience. Rather than meanings that exclusively reveal a "black experience," the songs sound the ultimate, most evocative translation of the universal human condition and the experiences common to all beings, all souls that live, suffer, and strive to overcome.

> "And so by fateful chance the Negro folk-song—the rhythmic cry of the slave—stands to-day not simply as the sole American music, but as the most beautiful expression of human experience born this side the seas. It has been neglected, it has been, and is half despised, and above all it has been persistently mistaken and misunderstood; but notwithstanding, it still remains as the singular spiritual heritage of the nation and the greatest gift of the Negro people." (Souls, 205)

Du Bois refers to the "group cry" of the Sorrow Songs also in *Dusk of Dawn.* "The colored world must be seen as a group whose insistent cry may yet become the warning which awakens the world to its truer self and its wider destiny."(Dusk, 679). This cry from within "the colored world" if heard not only offers the world the chance to open its eyes to a "truer" and "wider" destiny, it frees all who suffer from *white* -attitudes,—habits, and -wrongs. "The will to build a beautiful world- the quickest way to accomplish this is to listen to the complaint of those human beings today who are suffering most from white attitudes, from white habits, from the conscious and unconscious wrongs which white folks today are inflicting on their victims."(Dusk, 679). The "Cry" allows for an upheaval of ideals and images attached to the signifiers "black" and "white." Within the realm of the cry, black voices lead the way to beauty, truth and freedom, whereas "whiteness" distends oppressive, victimizing, attitudes, habits and wrongs.

Du Bois recounts and forewarns that most individuals listening to the songs on the "white side" of the Veil fail to appreciate their beauty, and the depth of meaning in the "cry." As Du Bois tells us the songs *are persistently mistaken and misunderstood* and the word "persistent" seems to suggest intentional misgivings concerning the songs' powerful messages. Surely, one reason people misunderstand the songs is due to their non-material quality. « Au-delà des objets désignés, au-delà des vérités intelligibles et formulées . . . il y a les essences, qui sont alogiques ou supra-logiques. » (Above designated objects, above intelligible and formulated truths . . . there are essences that are a-logical of supra-logical.) (*Proust et les Signes,* 80). The moment one tries to interpret the meaning of the songs in terms of pre-existing "material" formulas they inevitably loose their value. Deleuze tells us that one writes only for *illiterates;* for those who hear and absorb rather than "interpret and judge," for those who can appreciate the supra-logic of a-logical essences. (Dialogues, 90)

Donald Gibson who writes the book's introduction reads the value of the songs as non-value because he tries to interpret their meanings in terms of a world that makes its definitions of logic based materials. Although he notes the book's organic, non-logical quality, his reading typifies the negative, narrow response to the songs that Du Bois describes. In his thirty page introduction, Gibson reads then "writes" the songs off, in all of one paragraph, naming the songs "mute ciphers" that have "no meaning" unless one has had musical training and can "sight read." The introduction doesn't even discuss the entire chapter that dedicates itself to the songs and "leads the weary traveler" on his way to extend the lines, the sounds, beyond the book's borders. Gibson writes:

> "Though the double epigraphs at the beginning of the first thirteen chapters of Souls—the first a quotation from a well-known person or source, the second a musical notation of a phrase from various well-known black spirituals—are apparently intended to show the unity between the two modes of creativity, their actual effect is otherwise. . . . Even after we are told that the line of music beginning the first chapter is from "Nobody Knows the Trouble I've Seen," that has no meaning unless we know the song. . . . Those musical notations stand as mute ciphers, and rather than indicating, as Du Bois must have intended, that black people and white are in essence the same in that they possess souls, as attested to by the products of their creativity, the implicit message delivered is a grim one: the chasm that lies between black and white is as immense as the social, political, economic, and temporal chasm

> between Arthur Symons (the nineteenth-century British author of the first quotation and the creators of "Nobody Knows the Trouble I've Seen." (Souls, xvi)

Gibson states that he does not think that Du Bois intended the negative thought that even culturally black and white stand opposed facing an irreparable chasm. Nonetheless, he himself writes this bleak thought into the text's introduction. Gibson who is supposed to lead the reader into the text and make its meanings more accessible, instead shuts them down from the outset and leads *his* readers on a detour, off the textual tracks Du Bois traces and retraces. One instead finds themselves in a realm of Gibson's own presuppositions based on binaries that oppose music and poetic language, feeling and knowing, black and white, non-meaning and logical meaning. Gibson can't read correctly because the "verbal clatter" of his own mindset and that of the "major order" plugs up his senses and sensitivity. Gibson wrongly states that Du Bois fails to alert the reader to the fact that the musical notations are spirituals until the text's last chapter. However as already noted, Du Bois explicitly references the musical bars as "Sorrow Songs" in his own introduction. "Before each chapter, as now printed stands a bar of the Sorrow Songs,—some echo of haunting melody from the only American music which welled up from black souls in the dark past."(Souls, 2). Secondly Gibson's claims, clash cacophonously with Du Bois's text. Instead of these bars of music being "mute, meaningless ciphers," Du Bois tells us that they communicate the *cry*, the first hand history of black America and its version of American "democratic" practices. Du Bois incessantly emphasizes the importance, value, and meaningfulness, of the Sorrow Songs as well as their elusive, inarticulable quality that renders their significance deeper and more soulful.

> "They that walked in darkness sang songs in the olden days- Sorrow Songs- for they were weary at heart. And so before each thought that I have written in this book I have set a phrase, a haunting echo of these weird old songs in which the soul of the black slave spoke to men. The songs are indeed the siftings of centuries; the music is far more ancient than the words . . . knowing as little as our fathers what its words may mean, but knowing well the meaning of its music. . . . In these songs, I have said, the slave spoke to the world. Such a message is naturally veiled and half articulate. . . . Once in a while we catch a strange world of an unknown tongue, as the *Mighty Myo.*" (Souls, 204, 207, 209)

Du Bois reiterates time and time again, that these songs stand as the medium through which the slave could articulate in his own language

(Mighty *Myo*), a minor language, a message to the world that had to be "veiled, half-articulate." However, this does not make them a-significant. By naming the musical bars a "phrase" Du Bois elevates their signifying power and importance to the level of linguistic language.[9]

In light of all of this, the suggestion that one needs to read music or know the song, "Nobody Knows the Trouble I've Seen" first hand, in order for it to have any meaning is outrageous. One need only read the song's title to understand that its meanings communicate the unknown, unacknowledged sufferings and troubles of the outcast. In addition, Du Bois himself emphasizes that musical training is not a prerequisite to "understanding" the meaning of the songs. One who *hears* and excavates their *different, other-world* meanings, can register the other historical account that the songs relay. This account not only unmasks the socio-political hypocrisies inherent in American society but also advocates for change: for the generation of a truer world that institutes an all-embracing, truly just democracy.

> "What are these songs, and what do they mean? I know little of music and can say nothing in technical phrase, but I know something of men, and knowing them, I know that these songs are the articulate message of the slave to the world. . . . They are the music of an unhappy people, of the children of disappointment; they tell of death and suffering and unvoiced longing toward a truer world, of misty wanderings and hidden ways. . . . The world that are left to us are not without interest, and cleared of evident dross, they conceal much of real poetry and meaning beneath conventional theology and unmeaning rhapsody." (Souls, 207, 210)

Gibson and all *readers* who neglect the songs will necessarily create the "chasm" that Du Bois's text eliminates. Rather than "the British poet" Symons's work being dissimilar from the "sorrow song" that Du Bois pairs it with, his poem's words resonate intimately with it.[10] It seems strange for a literary critic to state that if one doesn't know the song it is meaningless as many literary works demand the reader's active participation, which often requires some background research. Usually when a quotation heads a prose piece the reader has the option to investigate that reference in order to understand its larger implications. Surely by 1989 when Gibson writes the introduction to *Souls* he could have accessed published and recorded versions of this classic spiritual. If he did he would have referred to its composer as Harry Thacker Burleigh rather than by the generic term "creators" and would have been able to aptly assess the similarities between "Nobody Knows the Trouble I've Seen," and Symons's poem. Both works speak of the despair, of feeling troubled. In addition Symon's poem which spearheads Du Bois's book relays the significance of the human cry, the major theme that runs throughout Du Bois's *Souls*. Symons uses the terms 'cry/crying' seven times in its short span of twelve lines, and 'voice' three times in the first four lines. "Voice of my heart, crying in the sand/ All night long crying with a mournful cry/ The voice of my heart in my side or the voice of the sea/. . . . And the heart shall be weary and wonder and cry like the sea." These lines that emphasize the cry or the voice could be those of the "Sorrow Songs" which Du Bois presents in his final chapter. If Gibson paid more careful attention to Du Bois's words, those of his text, and considered the textual selections more seriously, he would not contradict the entire direction of *Souls's* movements and curtail them. For Arthur Symons, a Welsh not *British* poet also wrote as a minority within the dominating control of the British Empire. Symons like Du Bois, also calls for a poetic expression that valorizes the non-material signs. This expression should sing, paint and evoke truths that come in non-conceptualized, "un-thinkable" forms. This singing sounds itself from a non-colored, non-classified, non-raced, *disembodied voice,* that of *the human soul.* In Symon's essay "The Decadent Movement in Literature" he expresses this: "It is the poetry of sensation, of evocation; poetry which paints as well as sings, and which paints as Whistler paints, seeming to think the colors and outlines upon the canvas, to think them only, and they are there . . . to be disembodied voice, and yet the voice of a human soul: that is the ideal of decadence." (http://homepages.nildram.co.uk/~simmers/symchron.htm)

In his introduction Gibson rather "reads," interprets and lends exclusive value to those aspects of the book that one can speak of and reduce to "historical," "social" and "cultural" realities. Gibson's reading "reterritorializes" the meanings that spring forth from the "middle" the in-between spaces, where connections are made between seemingly incongruous elements. Deleuze asserts that the rhizomatic text puts into play very different sign systems and even non-signs. . . . It is not made of unities but dimensions, or rather moving directions. There is no beginning or end, but always a (milieu) a middle, an environment. (Mille Plateaux, 31, 34, 36).

> "A rhizome-book, neither dichotomous, pivotal, nor fascicular. Never send down roots, or plant them, however difficult it may be to avoid reverting to the old procedures. . . . Let us summarize the principal characteristics of a rhizome: unlike trees or their roots, the rhizome connects any point to any other point, and its traits are not necessarily linked to traits of the same nature; it brings into play very different

regimes of signs, and even non-sign states. . . . It is composed not of unities but of dimensions, or rather moving directions." (Thousand, 23, 21)

« Un livre-rhizome, et non plus dichotome, pivotant ou fasciculé. Ne jamais faire racine, ni en planter, bien que ce soit difficile de ne pas retomber dans ces vieux procédés. . . . Résumons les caractères principaux d'un rhizome : à la différence des arbres ou de leurs racines, le rhizome connecte un point quelconque avec un autre point quelconque, et chacun de ses traits ne renvoie pas nécessairement à des traits de même nature, il met en jeu des régimes de signes très différents et même des états de non-signes. . . . Il n'est pas fait d'unités, mais de dimensions, ou plutôt de directions mouvantes. » (Mille Plateaux, 34, 31)

Paradoxically, it is the a-signifying aspects of the livre-rhizome that activate a politics of change but one must appreciate the "a-material" signs as a-logical or supra-logical. The moment one tries to "think" their meanings into logic, their meanings do indeed become muted. Deleuze writes in *Proust et les signes,* that the non-material signs, such as the musician Vinteuil's "petite phrase" or the Berma's expressions through the voice, gesticulation, and timber, open one to a world of an infinite possibility of *essences.* These non-material signs do so because unlike "material signs" they resist being interpreted in terms of already existing objects, conventional ideas and schemas.

"What is the superiority of Art's signs over all the others? This is because all the others are material. They are material first off due to their emission: They are halfway engrained (enveloped) in the object that carries them. . . . Without doubt Vinteuil's little phrases escape from the piano and from the violin . . . the Berma's expressions, like a grand violinist, have become qualities of timber. . . . Beyond designated objects, intelligible and formulated truths; but also beyond the subjective chains of associations and resurrections of resemblance and contiguity: there are essences that are alogical or supra-logical."

« Quelle est la supériorité des signes de l'Art sur tous les autres? C'est que tous les autres sont matériels. Ils sont matériels, d'abord par leur émission : ils sont à moitié engainés dans l'objet qui les porte. . . . Sans doute la petite phrase de Vinteuil s'échappe du piano et du violon . . . les expressions de la Berma, comme chez un grand violoniste, sont devenues des qualités de timbre. . . . Au-delà des objets désignés, au-delà

> des vérités intelligibles et formulées; mais aussi au-delà des chaînes d'associations subjectives et des résurrections par ressemblance ou contiguïté: il y a les essences qui sont alogiques ou supra-logiques. » (Proust et les signes, 47–49)

One can add, the musical notes, the strange, unique spirits that come through the voices of the Sorrow Songs, voices that sing in a plurality of languages including, African, German, and Afro-American vernacular, to Deleuze's references of non-material signs, that release truths that cannot be known through logic. (Souls, 207, 212). Proust writes, "Ideas formed by the intellect have only one logical and possible truth." « Les idées formées par l'intelligence pure n'ont qu'une vérité logique, une vérité possible. » (Proust et les signes, 22). In continuation of this idea Deleuze writes, "Ideas of the intellect are only worth their explicit signification, in other words one that is conventional. There are few themes that Proust insists on as much as this one: Truth is never produced by prerequisite good will, but rather results from a violence in thought."(22).

The reader of *Souls* must hear all elements of the text, and appreciate the heterogeneous plurality of signs it puts into play, being both material and immaterial. One must resist discounting those signs that break with logic and trigger a violent movement against conventions, instead of granting relevance only to those material signs that one can make "intelligent sense of." When Du Bois himself points to the elusive quality of his text he insists that if the non-material elements, the half-articulate, a-logical "counter-cries" that carry "burning truths" are not heard the life of the text falls "still-born." Du Bois explicitly announces this in his Cry, when he refers to the organic aspects of his book on its final page, with the signifiers; spring, leaves, vigor, reap, harvest, to underscore both the power and vulnerability inherent in such a discursive text.

> "HEAR MY CRY, O God the Reader, vouchsafe that this my book fall not still-born into the world-wilderness. Let there spring, Gentle One, from out its leaves vigor of thought and thoughtful deed to reap the harvest wonderful. (Let the ears of a guilty people tingle with truth, and seventy millions sign for the righteousness which exalteth nations, in this drear day when human brotherhood is a mockery and a snare.) Thus in Thy good time may infinite reason turn the tangle straight, and these crooked marks on a fragile leaf be not indeed THE END."(Souls, 217).

If the reader fails to recognize a text's pluralistic aspects than the reading is bound to eliminate and efface meanings and consequently shut the book down as the Veil shuts Du Bois out. If one does not *hear* the text with *infinite* reason rather than worldly logic, beyond what James names "verbal clatter," than one can not extend, and straighten the lines on the *leaves* to reap their harvest and sound their truths. However if one moves in the lines, and hears their "tingling truths," then they will not END, and those who suffer from a mockery of democracy may eventually be exalted.

Deleuze tells us that *lines of flight* are recaptured, reknotted, reterritorialized, in the molar or rigid segments of power centers that "seal, plug, and block" them, not because of misunderstanding, or oversight, but due to fear as Du Bois suggests in his comments on "persistent" neglect, and "despising." Deleuze states that we are always afraid of those elements that challenge the "great molar organization" that lends us a sense of stability and security.

> "We are always afraid of losing. Our security, the great molar organization that sustains us, the arborescences we cling to, the binary machines that give us a well-defined status, the resonances we enter into, the system of over-coding that dominates us- we desire all that. . . . We flee from flight, rigidify our segments, give ourselves over to binary logic; . . . we reterritorialize on anything available; the only segmentarity we know is molar. . . . The more rigid the segmentarity, the more reassuring it is for us." (Thousand Plateaus, 227)

What a reterritorializing reading misses are the "essences" or the "spiritual meanings" that the minor text proliferates through the singular worlds of "difference" that it generates. These d*ifferences* are overlooked because they are transmitted through discursive, non-formalized, non-material signifiers that can't be known or explained in the realm of logic or systems that encode. The differences activated by the non-material signs express a diversity of worlds that can only be revealed through art. The forces of these worlds of "difference" activate the kind of micro-political movements that this text referred to in its discussion of James's interpretation of the Dreyfus Affaire: Rather than directly rebelling against or contesting the validity of the Molar Unity, they make manifest counter, minor worlds that challenge and dismantle the foundations of unified systems. These worlds also invent and introduce new sonorities to deterritorialize the major language. Pure sonorous materials relay *other,* non-significant truths, as the minor writer *works* over the matter of the major language as Kafka mixes Czech and Yiddish into the German of Prague, and Black Americans make American English their language. (Kafka, 30, 37)[11] These *minor* works that reveal *minor* worlds produce "la petite musique," a way out of the *major* world and its language of oppression. "He was speaking of the "petite musique." Kafka as well, it is la petite musique, another, but always deterritorialized sounds. . . . Voilà, true "minor authors." A way out for language, music, writing . . . oppose the oppressed character of this language to its oppressive character, find the points of non-culture and under-development, the third world linguistic zones through which a language escapes itself." (Kafka, 49, 50)

THE VOYAGE BEYOND THE VEIL: DIFFERENT "I'S" WITH NEW "EYES"

> "An accent, Vinteuil's accent, separated from the accent of other musicians by a difference much larger than that which we perceive between the voices of two people . . . the question which he presented under so many forms, his habitual speculation, but also one unburdened, freed from the analytical forms of reasoning. . . . When the vision of the universe modifies itself, becomes no longer adequate for the meaning of the interior homeland, it is very natural that this then translates itself by way of a general alteration of the "sonorities" by the musician. The only "veritable" voyage, the only bath of "Jouvence," this wouldn't be to go towards new countrysides, but to have 'other' eyes, to see the universe with the eyes of an other, of one hundred others, to see the hundred universes that each of them saw, that each of them is; and that we can do with a Vinteuil, with those like him, we truly fly from star to star . . . I asked myself if music wasn't the unique example of that which could have been, the communion of souls, if there hadn't been the invention of language, the formation of words, the analysis of ideas." (Marcel Proust, *La Prisonnière,* 2, 3)

This chapter will now consider those invisible, muted, misinterpreted or overlooked elements, the *non-material signs* that activate an experience of "spiritual inter-subjectivity" and deterritorialize sounds, to write *ways out* of the oppressive, *great molar organization.* These neglected elements are the *souls,* sounds and stories of Du Bois's "*Black folk.*" In paying careful attention to the minor movements of these aspects of *Souls* one will hopefully appreciate how these elements along with the Sorrow Songs sound the sort of "accent" proliferated by Proust's ultimate musician "Vinteuil": An accent, unburdened from analytical forms of reasoning that modifies the vision of the universe. This vision-modifying accent proves the irreducible quality of souls, it leads its "listeners" on Proust's "only true voyage." This voyage does not concern a geographical displacement but instead provides news "eyes" through which to see the hundred universes that multitudes of other "I's" see. In other words, as one sees through other "eyes" *one* becomes "other." Subtly and surely the vision of the *whole* universe shifts. As Proust says, if one hadn't invented language, words and the analysis of ideas, wouldn't a realm in which souls could commune, interconnect, re-envision and create a truer ever-becoming universe exist? (La Prisonnière, 3).

W.E.B. Du Bois names his book, a composite of many genres, with the Sorrow Songs running through as consistent intertwining threads, *The Souls of Black Folk.* Considering once again the philosophy of William James wherein the particles ever create evolving, becoming wholes, along with the words of Gilles Deleuze that speak of the constant struggle between the verb Etre (to be/Being) and the conjunction "And"; the pragmatic way works to disrupt and replace notions of static "One" or "Everyone" groupings and "All" or "Whole" structures, as the "each ones" stimulate a becoming-everyone. The title of the book heralds Du Bois's, "each one" or anti–block-"every-one" structure, pragmatic approach.[12] Referring again to the title, straight away the message reads, "If society defines "Black Folk" as a unified homogeneous group, this text releases the heterogeneous plurality of '*Souls*' to undermine the notion of a reducible, one-dimensional, black identity. The book releases the "each one" mechanisms that attest to the varied, individualized, voices that have different universes to share within the rubric, "Black Folk."

Furthermore, Du Bois underlines that these 'identities' refuse to remain fixed and muted in the "Molar world's" categories. These souls live, and express their earthly plight. They struggle against the rules, orders, and prejudice that intern them in a sort of living, spiritual hell due to the Veil that shuts them off from the minimum requirements needed to feel a free and respected human being. "I pray you, then, receive my little book in all charity, studying my words with me . . . and seeking the grain of truth hidden there. I have sought to sketch, in vague, uncertain outline, the spiritual world in which ten thousand Americans live and strive."(Souls, 1). Du Bois provides the "eyes" through which to see these *thousands* of souls, their spiritual world, by inviting the reader as already stated, to the other side of the Veil. The Veil represents the barrier that shuts the black individual out of the white world. However this is not its only meaning, the Veil can also be interpreted as an apparatus that either perfects or impedes vision. It is alternately semi-opaque or translucent depending on which side of it one stands. One looks through the Veil and either sees elements and beings in distorted forms, or sees a world of cruel irony and injustice far too clearly. Du Bois seems to communicate that those on the *White* side of the Veil have a vision that is obscured by it and its corresponding ideologies that place it there in the first place, such as color prejudice and corresponding Jim Crow laws. Those on the other side, shut out from the larger world, "born with a veil" are to the contrary, "gifted with a second sight in this American world." (Souls, 5)[13] The negative side to the second sight is that the black individual lives conscious of the paradox of being shut out from a society that simultaneously groups them within a "democracy" that hails freedom and equality for all. "Even today the masses of the Negroes see all too clearly the anomalies of their position and the moral crookedness of yours. You may marshal strong indictments against them, but their counter-cries, lacking though they be in formal logic, have burning truths within them which you may not wholly ignore." (Souls, 88).

Gibson states that whereas most blacks must remain imprisoned within the Veil Du Bois moves freely from one side to the other and even rises above the Veil. Du Bois has the ability to move about it, step outside it, and even to lift it. (Souls, xii). "To some blacks it is a "prison house" whose walls are unscalable. . . . Du Bois, however, far from being held captive, has a sense of self and a self-possession so strong as to allow him to hold the world outside the veil in contempt and to live "above it in a region of blue sky and great wandering shadows." Gibson goes on to express uncertainty as to the meaning of "living above the veil." Gibson concludes that this must have to do with Du Bois's ability to compete successfully with whites. "Just exactly what Du Bois means by living above the veil is not entirely clear, yet it doubtless implies escape from the confines of the veil through the capacity to compete successfully with whites, with those who live outside it. The capacity to escape from the veil seems to be related to being outside the veil, and it also allows Du Bois to address the white world as he does in this book." (Souls, xii, xiii).

Gibson seems to suggest that if all blacks could just be as strong and intelligent as Du Bois and "successfully compete," than they too could live on all sides of the Veil. This in turn suggests that the color line is not an obstacle to a fulfilling life. This reading again in disagreement with Gibson says that Du Bois the man, does not move anywhere, but his narrator within the "world" of his book which establishes a new world order certainly does. In Du Bois's most personal, autobiographical chapter, "The Passing of the First-Born" that describes the painful loss of Du Bois's first-born infant underlines this. The only solace Du Bois can find in the face of his son's death is in the knowledge that his child will rise above the Veil. He will escape a destiny to live in a world that crushes ideals, spirits and lives. Du Bois on the other hand sees himself doomed to live as a captive of this world until his final days. "I shall die in my bonds." "Well sped, my boy, before the world had dubbed your ambition insolence, had held your ideals unattainable, and taught you to cringe and bow. Better far this nameless void that stops my life than a sea of sorrow for you. . . . Surely there shall yet dawn some mighty morning to lift the Veil and set the prisoned free. Not for me,—I shall die in my bonds." (Souls, 174). Du Bois the man, regardless of the fact that he is an intellectual, a scientist, a philosopher, remains fixed like all black American citizens behind the Veil identified as un-negotiably "Negro." The capacity to live above the Veil, to move oneself and others to different sides, to see and hear differently, "to escape" its bounds, comes from entering and co-creating in the realm where Culture reigns.[14]

> "I sit with Shakespeare and he winces not. Across the color line I move arm and arm with Balzac and Dumas, where smiling men and welcoming women glide in gilded halls. . . . I summon Aristotle and Aurelius and what soul I will, and they come all graciously with no scorn nor condescension. So, wed the Truth, I dwell above the Veil. . . . This, then, is the end of striving: to be a co-worker in the kingdom of culture, to husband and use his best powers and his latent genius." (Souls, 90)

Du Bois works to co-produce the *Kingdom of Culture* throughout the book. The short story, "The Coming of John," may be seen as the most concentrated expression of this effort. However, while *The Souls of Black Folk* has been acknowledged as a key contribution in the creation of the "Kingdom of Culture," *The Coming of John* has been primarily read as a less significant part of a more significant whole. Du Bois included the short story in the compilation of writings known as *Souls,* not so it would go unheeded but rather to ensure that it would be published.[15] Not only would it be published, but "hidden" in a book which presented itself as a sociological, historical study of the black American population, also increased its chances of being read by a larger audience. However, because Du Bois had to situate his story in such a rich text, it has most often received little more than a passing, superficial commentary.

Ralph Ellison's corpus of critical non-fiction bears testimony to this fact. Ellison states that with the exception of William Faulkner, most twentieth century writing renders a vacuous, de-humanized portrayal of the Negro figure whereas the great American writers of the 19th Century such as Melville, Thoreau, Emerson, and Whitman revealed the Negro as the ultimate symbol of Man. (Shadow, 32, 43). "Thus when the white American, holding up most twentieth-century fiction, says, 'This is American reality,' the Negro tends to answer. 'Perhaps, but you've left out this, and this, and this. And most of all, what you'd have the world accept as *me* isn't even human.'" (Shadow, 25).

Ralph Ellison did not fail to mention Du Bois's story as one that indeed creates the kind of complex non-stereotypical, representations of black subjectivity he calls for in his essays, because he was unfamiliar with Du Bois's work. Ellison who often uses the Du Boisean trope of the Veil, in both its negative and positive sense, as a barrier to a fuller non-colored life, and as the caul that lends "second sight" (Shadow, 29, 61), indicates the degree to which *Souls* influenced his own thinking and writing. Like most critics, Ellison overlooks *The Coming of John* as an exemplary model of the kind of variant "democratic" writing he speaks of that produces "other" representations of American life, because one thinks of it and reads it as one chapter among many, rather than as a short story. (Shadow 43, 44, 26). Du Bois's story written in 1903, gives textual life to a black male protagonist that the reader views through a distinct, and vision adjusting conceptual lens that projects John's interior homeland. This lens portrays John and his interpretation of the various worlds that surround and *keep him in his place,* from within the black zone of the Veil. One also hears the sounds that resonate in John's ears with an auditory filter of an *other* world sensitivity. Again Deleuze tells us in *Proust et les Signes* that only non-material signs, signs exclusive to Art, communicate spiritual truths. These spiritual truths emitted in Kafka's "cry," Vinteuil's "petite phrase," La Berma's voice, generate a "Difference" that allows us to see the diversity of worlds that the unique "spiritual" essences reveal.

> "Art gives us a veritable unity: unity of an immaterial sign and of an all spiritual meaning. . . . Essences or Ideas, this is what each sign of the petite phrase reveals. What is an Essence which is revealed in a work of art? This is a difference, the ultimate and absolute Difference. It is this which constitutes being and helps us conceive being. This is why art, to the degree that is manifests essences, is only capable of giving us what we seek in vain in life. . . . *Only through art, can we escape Us, come to know that which another of this universe sees, which is not the same as our own. Thanks to art, instead of seeing a sole world, ours, we see it multiply itself, and to the extent that there are original artists, we have worlds at our disposition, so different one from the other as that which rolls on in the infinite.*" (Proust, 47–49)

> « L'Art nous donne la véritable unité: unité d'un signe immatériel et d'un sens tout spirituel. . . . Des essences ou des Idées, voilà ce qui dévoile chaque signe de la petite phrase. Qu'est-ce qu'une essence, telle qu'elle est réveillée dans l'œuvre d'art? C'est une différence, la Différence ultime et absolue. C'est elle qui constitue l'être que nous fait concevoir l'être. C'est pourquoi l'art, en tant qu'il manifeste les essences, est seul capable de nous donner ce que nous cherchions en vain dans la vie. . . . *Par l'art seulement, nous pouvons sortir de nous, savoir ce que voit un autre de cet univers qui n'est pas le même que le nôtre. Grâce à l'art, au lieu de voir un seul monde, le nôtre, nous le voyons se multiplier, et autant qu'il y a d'artistes originaux, autant nous avons de mondes à notre disposition, plus différents les uns des autres que ce qui roulent dans l'infini.* » (*Proust et les signes*, 52, 53)

The Coming of John releases the kind of non-material signs through its "wail" that like Vinteuil's "petite phrase" and Kafka's "cry," reveals an *all spiritual meaning.* One of the ultimate "Differences" that reveals an *other* "universe," which distinct from our own, challenges our perception of the "One world" as we think we know it.

Summarized in terms of its events, the story might be read as a cliché representation of the problem of race prejudice in America: The main character, black John and his white, boyhood friend, "white John," grow up and leave their small, southern town to go to college. The boys return after many years, educated and worldly. Black John hopes to teach the black community and move them towards social uplift. Neither the black nor white world can accept the transformation of Black John. The black world no longer understands *his language* and the white world fears the changes he might incite in the black group. The Judge of the town finally casts Black John out of the broken down Negro school where he returns to teach, when the postmaster reports, "Heah that John is livenin' things up at the darky school, ohm nothin' in particulah,—just his almighty air and uppish ways. B'lieve I did heah somethin' about his givin' talks on the French Revolution, equality, and such like. He's what I call a dangerous Nigger." (Souls, 199) On his way home from this violent and painful dismissal, Black John unintentionally kills the Judge's son, White John with a tree limb when he sees his sister struggling with the latter as she tries to resist sexual assault. Black John finally awaits his death as the Judge and his posse ride to lynch him.

The story line is not complex but the mixture and merging of diverse and disparate elements generates the kind of singular "Difference" Deleuze refers to. That Difference that Deleuze tells us "constitutes the being that makes us conceive being." The being is not already there as part of some pre-existing material but is born of a new, non-material expression. The difference that evokes the "spiritual" truth, that of individual souls results from an artistic style that takes two seemingly opposed "objects" and expresses a quality common to both of them that one comes to perceive in the *milieu révélateur* (revealing middle).

> "'*Truth does not begin until the moment the writer takes two different objects, poses their relation, analogue in the world of art . . . and encloses them in the necessary links/rings of a beautiful style.*' . . . This is to say that the style is essentially that of metaphor. But the method is essentially metamorphoses, and indicates how the two objects exchange their determinations, exchange even the names that designate them, in the new milieu that confers them their common quality. Such as in the paintings of Elstir, where the sea becomes the earth, and the earth, sea, where the town is only designated by 'marine terms,' and the water, by 'urban terms.'"

> « « *La vérité ne commencera qu'au moment où l'écrivain prendra deux objets différents, posera leur rapport, analogue dans le monde de l'art . . . et les enfermera dans les anneaux nécessaires d'un beau style.* » C'est dire que le style est essentiellement métaphore. Mais la métaphore est essentiellement métamorphose, et indique comment les deux objets échangent leurs déterminations, échange même le nom qui les désigne, dans le milieu nouveau qui leur confère la qualité commune. Ainsi dans les tableaux d'Elstir, où la mer devient terre, la terre, mer, où la ville n'est désignée que par les « termes marins, » et l'eau, par des « termes urbains. » » (Proust et les signes, 59)

In Du Bois's text one can see a number of metamorphoses activated through the particular style that brings two different elements together in one metaphor. One has a multiplicity of becoming Johns, becoming worlds through metaphors which subsequently associate, Black John with John the Baptist, Black John with the American ideal of the self made-man, Black John with the white Swan, and Black John with the Swan's song. The title of the story, "*The Coming of John*" along with the theme of the Cry that runs throughout *Souls* signals a metaphoric relationship between black John and the "Precursor," to Jesus, John the Baptist. When the Jerusalem priests and Levites ask John the Baptist if he is the savior he responds, "I am the voice of one crying in the wilderness, make straight the way of the Lord, as said the prophet Isaiah." (John, I, 19–23). Isaiah reads, "The people who walk in darkness have seen a great light; those who dwell in the land of shadows of death, upon them the light has shined."(Isaiah, 9). Both of these biblical passages echo several phrases in Du Bois's text. The chapter, "The Sorrow Songs" which echoes "Isaiah" begins: "They that walked in darkness sang songs in the olden days- Sorrow Songs- for they were weary at heart." (Souls, 204). Just thirteen pages later Du Bois urges the reader to embrace the meanings, the life of his book with "infinite reason" so that it does not fall still-born in the "world-wilderness," using language very similar to that of (John, 1) just cited: "HEAR MY CRY, O God the Reader; vouchsafe that this book fall not still-born into the world-wilderness. . . . Thus in Thy good time may infinite reason turn the tangle straight." (Souls, 217). Black John, as John the Baptist again prepares the world to *straighten the way,* to hear and live by *futurizing,* power destabilizing messages because Black Americans never experienced liberation from spiritual and moral slavery on earth due to the white world's refusal to fully *hear* the "cry," the message of "John." Black John voices the Cry in the wilderness again, that cry that Du Bois's narrator also sounds, to emphasize the need for the tangles, the paths, to *be made straight,* to be extended spiritually and psychically.

Another metaphor in the story presents John as America's self-made man. John who begins his student-life as a boisterous, clumsy, tardy class-clown that is asked to leave school, willfully returns a different man. Through his own determination to finish school and make his community proud John changes his habits and changes his person from what the black is supposed to be, a static, fixed "being," to rather exemplify the activity of a *becoming*-being.

> "When he came back to us he went to work with all his rugged strength. It was a hard struggle, for things did not come easily to him,—few crowding memories of early life and teaching came to help him on his new way; but all the world toward which he strove was of his own building, and he builded slow and hard. As the light dawned lingeringly on his new creations, he sat rapt and silent before the vision, or wandered alone over the green campus peering through and beyond the world of men into a world of thought. . . . So he thought and puzzled perplexed where others skipped merrily, and walking steadily through the difficulties where the rest stopped and surrendered." (Souls, 190)

John wills himself to become a thinker and undergoes an intellectual and spiritual, metamorphosis. In thoughtfulness John becomes a complex, grave, well-spoken sage, in sum, an exceptional man. John builds his own pluralist world as he confronts the paradoxes and the difficulties that the average person retreats from: This world allows him to build unique creations and lends him new visions. John like Du Bois's narrator soars above the Veil when he takes up his "new way," or way out, "beyond" the racially polarized world he grew up in. John thinks and feels above the mean, above the "real" status quo world-order. However, when John graduates from the University suddenly he faces the less-than, narrow world that subjugates him to a "low life."

> "He had left his queer thought-world and come back to a world of motion and men. He looked now for the first time sharply about him, and wondered he had seen so little before. He grew slowly to feel almost for the first time the Veil that lay between him and the white world; he first noticed the oppression that had not seemed oppression before. . . . He felt angry now when men did not call him "Mister" he clenched his hands at the "Jim Crow" cars, and chafed at the color-line that hemmed in him and his. A tinge of sarcasm crept into his speech, and a vague bitterness into his life; and he sat long hours wondering and planning a way around these crooked things." (Souls, 191)

When John goes to New York one finds him again encapsulated in another metaphor. This metaphor associates the wail of Wagner's Swan, in the opera "Lohengrin," with John's "Cry." The wail expresses the bitter angst John, *the symbol of man* feels when he confronts the cruel reality of the "crooked" oppressive ways and codes of the Jim Crow world that makes him travel on the back cars, enter at back doors and be treated as less than a man. When John travels north to New York to sing with his school's quartet, a crowd sweeps him along with it and he finds himself attending Wagner's *Lohengrin.* The power of the swan's wail overwhelms John and permeates his being. "The infinite beauty of the wail lingered and swept through every muscle of his frame, and put it all a-tune. He closed his eyes and grasped the elbows of the chair, touching unwittingly the lady's arm. . . . A deep longing swelled in all his heart to rise with that clear music out of the dirt and dust of that low life that held him prisoned and befouled. he felt with the music the movement of power within him." (Souls, 193). Simultaneously the beauty of the swan's wail makes John aware of the "movement of power within" his own person. The wail of the swan translates his experience of the world perfectly, "putting it all a tune." In other words the wail that sounds itself from an-other world makes more sense of John's existence than those meanings generated by the laws, boundaries, narrow schemas, of the white world and its attempt to define him and hem him in. The significance of the Wail parallels Deleuze's description of Kafka's cry released by the magistrate.

> "'It seemed not to come from a man, but rather from a machine of suffering.' . . . In sum, the sound doesn't appear as a form of expression here, but as a *non-formed matter of expression,* which will react upon other terms. . . . That which interests Kafka, is a pure, intense, sonorous matter . . . musical sounds deterritorialize, cries that escape signification . . . sonority ruptures in order to free itself from a chain which remains too significant. In the sound, only the intensity counts, generally monotone, always a-signifying."
>
> « Il ne semblait pas venir d'un homme, mais d'une machine à souffrir » . . . Bref, le son n'apparaît pas ici comme une forme d'expression, mais bien comme une *matière non formée d'expression,* qui va réagir sur les autres termes. . . . Ce qui intéresse Kafka, c'est une pure matière sonore intense . . . son musical déterritorialisé, cri qui échappe à la signification . . . sonorité rupture pour se dégager d'une chaîne encore trop signifiante. Dans le son, seule compte l'intensité, généralement monotone, toujours asignifiante. » (Kafka, 12)

The sound of the Wail, like the "cry" of Kafka, Du Bois, and John, is a-signifying if one tries to translate it into language because it is non-linguistic and generates *minor* meanings. The Wail and the cry deterritorialize as they express the inexpressible, meanings, truths, realities muted in the major world order. The sound of the wail that articulates John's world, imagination, his inner strength in movement, carries him to a place above the "dirt and dust of the low life." The swan in the opera also mirrors John's situation of "imprisonment": For the swan is really Duke Godfrey who Ortud viciously curses into animal form. Godfrey like John lives confined to a body that causes others to view him as non-man, as animal as less than human. In this body Godfrey like John has no language with which to confront his state of internment. Like Du Bois, Godfrey, and Du Bois's *Souls,* John can only articulate his reality through a "cry."

As already stated the difference that reveals the all "spiritual," immaterial sign or essence occurs as metaphor becomes metamorphoses, as each different element works one on the other to transform each one into something other. Through the juxtaposition of Black John and the White Swan, Du Bois communicates that the entities of the Wail hold more in common than the superficial aspects that define them as separate. Taken in by the wail of the swan, John becomes a nobleman entrapped by what is perceived as a beast's body. The swan similarly becomes a captive of cruel fate, arbitrary powers and laws that render a man less than a man, force him to suffer, to feel trapped, to cry out a-significant deterritorializing sounds.

The story's final scene makes the metamorphoses more complete. After Black John has killed White John he waits to be lynched in the calm spirit of the noble martyr. The wail of the swan and Wagner's music envelops John and releases his spirit to an enlightened realm of being, that transports him beyond the surrounding, mundane brutality, symbolized by the act of lynching. In this realm, John righteously ascends to a position where he views and defines the "haggard, white-haired judge" and the world he represents; its "justice system," power and order, as nothing more than "pitiful."

> "Then as the sheen of the starlight stole over him, he thought of the gilded ceiling of that vast concert hall, and *heard* stealing toward him the faint sweet music of the swan. Hark! Was it music or the hurry of shouting men?. . . . He leaned back and smiled toward the sea, whence rose the strange melody. . . . With an effort he roused himself, bent forward, and looked steadily down the pathway, softly humming the "Song of the Bride,"- "Freudig gefuhrt, ziehet dahnin." Amid the trees in the dim morning twilight he watched their shadows dancing and heard their horses thundering toward him, until at last they came sweeping like a storm, and he saw in front that haggard white-haired man. . . . Oh, how he pitied him,—pitied him. . . . Then, as the storm burst round him, he rose slowly to his feet and turned his closed eyes toward the Sea. And the world whistled in his ears." (Souls, 203)

Within this final overture one hears, feels and enters John's world that merges with "The Kingdom of Culture." In this New World John sings the German words to Wagner's, "The Song of the Bride." The notes to *Souls* tell us that Du Bois changes the German line from "led faithfully" to "led happily." (Souls, 244). Through the music, John finds his "way around these crooked things," the pathway that leads him *happily* forward, beyond the white world's limits and laws. The music of the story, like the "cry" of Du Bois's *Souls,* activates what Deleuze calls "les lignes de fuites." Lines of escape or leakage, that puncture and cause leaks in, or ways out of the system, or as Deleuze states that "font fuire le system" (that makes the system take flight/off). The *Sorrow Songs,* the lines of poetic verse, Wagner's opera, all co-create the "Kingdom of Culture." In this realm there are only disembodied voices, there are no colors, races, or genders there are only souls striving to activate becomings, and make the world a becoming world. The voice that sings the spiritual that bespeaks a broken heart is the same voice that sings the German folk song. (Souls, 212).[16] The voice that releases the Wail is the voice of Godfrey and Black John, that voice that cries in the sand without avail is the voice of Arthur Symons, John the Baptist, and "Black John." The voice that calls for the preparation of a new Kingdom on earth where all may live to fulfill their ultimate potentials is the voice of W.E.B. Du Bois, and the chorus of the multitude of heterogeneous "Souls" of the sorrow songs, past, present and future. The world that allows for a proliferation of different voices, of different worlds, of "non-material," *spiritual* truths on earth is the "Kingdom of Culture." Within the Kingdom of Culture new truths can forge new paths, new territories and deterritorialize old ones. In this realm one reaches the ultimate political passage. That of James's microcosm of effects that alters sonorities, grants new visions and expands the world's notion of itself so that it may include the oxymorons, the unclassifiable residuums and esteem those formerly seen as problems as *other* world prophets and saviors.

Chapter Three

Traitors versus Cheaters: *Le Devenir Imperceptible* in the Writings of Gertrude Stein, Jean Toomer, and William Faulkner

> "Truth always has "jagged edges." . . . not a uniform piece of clothing but a patchwork, with multiple joinings . . . the American invention *par excellence*. . . . But to reach this point, it was necessary for the knowing subject, the sole proprietor, to give way to a community of explorers, the brothers of the archipelago, who replace knowledge with belief, or rather with "confidence." . . . Pragmatism is this double principle of archipelago and hope. (These themes are to be found throughout Pragmatism, and notably among William James's most beautiful pages: the world as "shot point blank with a pistol." This is inseparable from the search for a new human community." (Bartleby, ou la formule)

As Deleuze tells us, a livre-rhizome or a rhizomatic writing or thought always creates a milieu. The milieu as in-between operates as a generative unit. As Wahl tells us of James's "relations," the "avec"(with), one can imagine a world of the "avec" the "milieu" and nothing but because they are so fecund they generate infinite off shoots, that extend in a plurality of directions. These "avec," the radical imperceptible, molecular particles, hold multiplicities as they connect with diverse elements and then again break apart as different again, to subtly, silently, but persistently push the whole towards a constant destabilization, restructuring and mutation. Du Bois's *Souls of Black Folk* as the last chapter argues and hopefully demonstrates, definitely activates the "immaterial" signs, particles, sounds, music, the unclassified residuum that the unifying, negating, totalitarian systems leave out, fail to see, to hear, or push out of their peripheries.

As already mentioned, minor literature or "les livres-rhizomes" activate collective *agencements.* As the writer undergoes a "devenir imperceptible," he/she also deterritorializes the "major" standardized, language, and activates a series of events that shift, and destabilize the foundations of the "molar system," to realize a "devenir révolutionnaire." Deleuze insists that to create in such a way demands a betrayal by the writer of his/her class, race, sex, or majority. One will pursue several questions in this chapter as it considers these aspects of *le livre rhizome* with regards to the distinction between the writer that *betrays* and the writer that *cheats.* The first text concerned like Du Bois's *Souls,* owes something to James's pragmatic pluralism. This is Gertrude Stein's work "*Melanctha*" that Stein even dedicates to William James. (Autobiography, 80). The next three works read are Jean Toomer's *Cane,* Sherwood Anderson's *Dark Laughter* and William Faulkner's *The Sound and the Fury.* These texts either directly or indirectly stem from one the other or both of the first two books mentioned. In relation to Deleuze's distinction between the writing of "the traitor" verses that of "the cheater," this text will ask, which books betray and why? What is the difference between a minority and a becoming-minoritaire? How does a becoming-minoritaire, becoming-woman, becoming-black, becoming-imperceptible activate a becoming-revolutionary? How in the end do these books that betray indeed affect a micro-politics, a "pragmatics"?

As the final pages of the first chapter relay, Deleuze's prescription for the literary artist's "success" stands diametrically opposed to that of Sartre. Sartre's writer must write from the situation and vantage point of his/her "historicity" and social position, and improve the lot of the group with whom he/she is associated. Deleuze's writer to the contrary, must "betray" his/her "majority," social definition, assigned identity in other words to use Deleuze's own expression, "take flight and at the same time, puncture, "faire fuir" the system"(Chase away the system).

In *Dialogues* Deleuze writes that the individual is always "pinned against the wall of dominant significations" that is inscribed with "objective determinations" that fix and identify us. One gets stuck upon this "grille," in the *black hole* of subjectivity, of the Ego. (Dialogues II, 45). Deleuze writes that to get oneself out of the mechanisms that "pin one to the wall," one must lose face, which is a prerequisite for "betraying." For Deleuze, the face operates as a symbol of the socially identifiable and identified subject or social object. "*The face is a product of this system, it is a social production.*"(45). One must "loose face" in order to generate ways out, "les lignes de fuites." Again he associates this activity with a capacity to "love"; blindly, experientially, in a realm beyond bodies, categories, memory (history), and interpretation to attain the ultimate creative venture: "experimentation-life." The writer must become clandestine, unknown, and enter various becomings, -animal, -Negro, -woman. Gilles Deleuze insists that this "betrayal" entails true artistry: "It is very difficult to not be known by one's landlord, or in one's neighborhood, the nameless singer, the ritornello." (Dialogues II, 45)

In becoming- "others," -"*minoritaire,*" one traces out events, experiences, and "third world or minor world linguistic zones," voices, timbres, accents, musical notes or sounds that the major linguistic grid will not or can not hold.[1] As somewhat touched upon in the last chapter; minor authors, the greatest authors, find their way out, jump over, traverse and simultaneously pierce the wall that tries to bind and contain the individual within fixed identities by "conquering" the "major language."

> "That is the strength of authors termed "minor," who are in fact the greatest, the only greats: having to conquer one's own language. . . . It is in one's own language that one is bilingual or multilingual. . . . Use the minor language to send the major language fleeing." (Thousand, 104, 105)

> « Telle est la force des auteurs qu'on appelle « mineurs, » et qui sont les plus grands, les seuls grands: avoir à conquérir leur propre langue. . . . C'est dans sa propre langue qu'on est bilingue ou multilingue . . . . Se servir de la langue majeure pour faire filer la langue majeure. » (Mille Plateaux, 132, 133)

Deleuze uses the example of "black-English" as a minor language that invades the "major language" and puts it in a state of continual variation, but insists that it is not necessarily the identifiable "black subject" that writes "black American language" into literature to stimulate a "devenir-minoritaire" or a "devenir-noir." In addition, this devenir has nothing to do with "imitation" or "mimicry" and everything to do with what Deleuze describes as a mutual becoming of two distinct entities. Deleuze explicitly addresses this misconception in *Dialogues.* He clearly states that it is not necessarily the female writer that activates a devenir-femme, a black writer that activates a devenir-noir, the only essential ingredient is that the writer 'betrays' rather than 'cheats' to install this "double capture between two reigns" wherein each deterritorializes the other. Pointing to the examples of D.H. Lawrence and Henry Miller, Deleuze explicates this. Although most consider them "great phallocrats," their writings draw them into "irresistible woman-becomings." Deleuze states, a *Woman* is not necessarily the writer, but the becoming-minoritarian of the writing, whether it be by a woman or man. "What other reason to write than to be a traitor to one's own reign, traitor to one's sex, to one's class, to one's majority? And to be traitor to writing." (Dialogues II, 43, 44).

Perhaps Gertrude Stein above all others most remarkably exemplifies the "traitor" who writes and lives to "betray" her sex, language, *race,* class. Stein, a lesbian, expatriate, Jew, similar to Du Bois renounces the idea of having one fixed, stable identity and criticizes the American cultural, social and political milieu that she found to be very undemocratic and stifling. In fact Lynn Weiss who writes, *Gertrude Stein and Richard Wright* states that Stein brings up the paradox that her originality in life and in art relates to having been raised in the margins of a nation that, still becoming could not really "identify" what she was. Weiss goes on to point out that in *Everybody's Autobiography,* Stein refutes nouns that encourage essentialist perspectives such as; "Jew," "woman," "Californian," and "lesbian." "*Stein's chief criticism of America was its denial of process and its wholehearted investment in the meaningfulness of these categories.*"(23). Stein who quits medical school in the United States to set up house with her lover Alice B. Toklas in Paris, where she becomes the matriarch and salon guru of the left banks art's scene also studied with William James and repeatedly refers to his influence in her development and thought. Again this tutelage encourages a constant skepticism and questioning of "givens." In *The Autobiography of Alice B. Toklas* Stein indicates that James's introduces her to the kind of thought that entices her while it promotes *process,* pleasing amusement and considers absolute "truth" abject.

> "The important person in Gertrude Stein's Radcliffe life was William James. . . . She was the secretary of the philosophical club and amused herself with all kinds of people. She liked making sport of question asking and she liked equally answering them. She liked it all. But the really lasting impression of her Radcliff life came through William James. . . . William James delighted her. His personality and his teaching and his way of amusing himself with himself and his students all pleased her. Keep your mind open, he used to say, and when someone objected, but Professor James, this that I say, is true. Yes, said James, it is abjectly true." (Autobiography, 78, 79)

The work that Stein associates above all with William James's influence, teaching, and open thought is her publication *Three Lives.* Not surprisingly, it is in this work that Stein truly "betrays" and generates innovative, literary techniques and stylistic devices that achieve a "devenir minoritaire" that writes against stabilizing notions of both gender and race. Again in the

Autobiography Stein communicates that James's himself identifies his teaching as instrumental in her ability to make her "openness" manifest. According to Stein, when James visits her in Paris and looks at her "pictures" and "writings" he gasps and responds, "I always told you that you should keep your mind open."(Autobiography, 80). In the paragraph that follows, Stein reveals that she dedicated her *Three Lives* to William James and sent him a copy. She again emphasizes that James found it of such interest that he did not just read it but felt compelled to write with it, making notes in its margins. Stein sets the scene recounting that a lawyer from Boston contacted her with news that he had the copy that she had sent William James (now deceased).

> "He said in his letter that he had not long ago in reading in the Harvard library found that the library of William James had been given as a gift to the Harvard Library. Among these books was the copy of *Three Lives* that Gertrude Stein had dedicated and sent to James. Also on the margins of the book were notes that William James had evidently made when reading the book. The man then went on to say that very likely Gertrude Stein would be very interested in the notes and he proposed, if she wished, to copy them out for her." (Autobiography, 80) [2]

The section of *Three Lives* that Gertrude Stein most emphatically promotes that actually reads more like a short novel than a story, is *Melanctha.* Stein in fact names this work one of the primary sources that helps her "kill" the nineteenth century. "Stein asserted that she had "killed" the nineteenth century much as a "gangster" takes out his victim; she was, she said, the twentieth century."(Terrible Honesty, 120). Stein refers to *Melanctha,* which she wrote in black vernacular many times in *The Autobiography* as that work which gives her an advance on the new century. "Gertrude Stein had written the story of Melanctha the Negress, the second story of *Three Lives* which was the first definite step away from the nineteenth century and into the twentieth century in literature."(Autobiography, 54 /82).

Stein's writing is truly curious and unconventional, even amidst modern models. One enters a narrative that spirals around and around but never goes anywhere in time or place. The text jockeys the reader back and forth. One enters a realm of ebb, flux, flow and refrain devoid of a sense of development or progression. In fact Stein's narration repeats phrases, descriptions, allusions so frequently that one has the final sense that in the span of 107 pages, Stein may have set the record for the fewest number of different words used. The narration disorients the reader as it leads him/her into unfamiliar contexts of narrative, language, meaning/association, and time. Stein's New York publisher of the Grafton Press assumed that she wrote as a foreigner writing in English due to her strange, language usage. In addition, rather than telling a tale with a beginning, and end, that focuses on sequential events that lead to the some final result or moral, she creates a milieu that allows for a number of becomings to potentially take place. This kind of writing as *act of thought* blazes new trails and sows seeds that stretch out and surge up from the middle. Due to the particular style of *Melanctha* Stein generates characters that one can compare to those of Becket's writings that Deleuze describes: These characters always in the middle perpetually involute. Contrary to individuals that one expects to grow according to genetically predictable paths and structural organizing codes these becoming-individuals take the unpredictable path of the rhizome that moves free of past or future orientations. Deleuze writes that because American writers such as Stein, create in this way they annulled the need for philosophy as a "specialized institution" because their novels made writing "an act of thought," and life a "non-personal power."

Along with being stylistically unique and innovative particularly for its time, the text contextually also breaks with conventional subject matter, characterizations and plot structure. The story about a young bi-racial girl includes only "black" characters. As in the case of Du Bois's text the "majority" reader enters a world that they are not altogether familiar with. One reader of *Melanctha,* Miss Godard King comments that the sense of strangeness Stein's text evokes is due to the fact that it leads the reader to a "colored" milieu wherein the reader momentarily undergoes a "devenir-colored." "The third case is strange . . . emerging in the life of a young colored girl. And in what reality! Everyone in the story is colored, the whole world, with all its preoccupations, and potentialities; the reader himself, for the time, is a colored person too. . . . The patient iteration, the odd style, with all its stops and starts, like a stubborn phonograph, are all a part of the incantation."(xviii). The oddity of Melanctha's "colored world" is that it lures one into a becoming-black that is not racial. What does that mean? Stein's text divorces her "colored" world from the white, Major World order and lends autonomy to its languages, codes, exchanges, operations, and characters. Therefore Stein's all "black" cast's "blackness," which normally would be emphasized through signs of racial difference, fades away.[3] Furthermore, the story only describes two of the characters as *black.* Melanctha is both described as, 'half white,' and along with her mother, as 'pale yellow.' Other characters are described as 'pale brown,' or 'light brown.' Stein at the same time undermines essentialist claims that associate certain actions or qualities with "blackness" or color specificity. The only "black" characters, Rose Johnson and Melanctha's father James Herbert, fail to laugh as "real" Negroes, the latter two, "never had the wide abandoned laughter that gives the broad glow to the Negro sunshine."(Lives, 59, 64). One can perhaps better understand this fading away of "blackness" as racial, in terms of Deleuze's description of *race,* as something arbitrarily in place while an exterior order designates an "other' group as inferior. "The race-tribe only exists on the level of an oppressed race, and in the name of oppression that it suffers: there is no race but inferior, minoritarian; there is not any dominant race, a race is defined not by its purity, but on the contrary by the impurity conferred upon it by a system of domination." (Thousand Plateaus, 370). Because Stein's text removes the "micro-political, black world" with its "micro-political language" from the "binary machine" that defines it only in relation to white customs, laws and codes, she achieves a *de-race-i-nation.* Due to this, one is no longer truly in a colored world but a world becoming-colored or perhaps more colorful or more sonorous. Miss Godard King who speaks of the text's musical, phonographic quality makes a very astute comment, for Stein herself compares Melanctha's composition to that of Bach's musical fugue.

> "During these long poses and these long walks Gertrude Stein meditated and made sentences. She was then in the middle of her Negro story Melanctha Herbert, the second story of *Three Lives* and the poignant incidents that she wove into the life of Melanctha were often these she noticed in walking down the hill from the rue Ravignan. . . . She had come to like posing, the long still hours followed by a long dark walk intensified the concentration with which she was creating her sentences . . . by refusal of the use of the subconscious Gertrude Stein achieves a symmetry which has a close analogy to the symmetry of the musical fugue of Bach." (Autobiography, 80)

This passage also reveals that Melanctha although situated in the context of "colored life," transmitted through the sounds of a Steinified black vernacular is not an attempt to present a reductive portrayal of a sort of ghettoized black life. Stein's text evokes life that is worldly, cosmopolitan in scope. This life interweaves elements of human exchanges that Stein hears and sees taking place in Paris on the Rue Ravignan, with those sounds and scenes she recalls as a medical student at John Hopkins engaged in field work. "It was then that she had to take her turn in the delivery of babies and it was at that time that she noticed the Negroes and the places that she afterwards used in the second of the *Three Lives* stories, Melanctha Herbert, the story was the beginning of her *revolutionary work.*"(Autobiography, 54 /82). Stein literally and figuratively creates *Melanctha* in an *in between* space. *Melanctha* the most important story is situated in the middle of her two other stories and Stein fabricates it out of the milieu that combines two different perspectives of lives that live on two different national and cultural terrains. What may be the most ground breaking aspect of Stein's work is that as Stein betrays writing, notions of race, and gender, she allows for a series of devenirs that the reader enters simultaneously. *Melanctha* indeed activates a de-racinated, devenir-noir but the absence of "whiteness," its laws, and representative of the patriarchy, being the white male, that keeps blacks and women in their proper places, allows for an even more intense devenir-woman.

The introduction to *Melanctha* written by Ann Charters, underlines this "feminine" aspect of Stein's text. "We are in a woman's world in *Three Lives,* one in which women's voices are heard expressing their inner feelings, wishes, moods, ideas, and confusions and above all else, earnestly advising one another of the best way to live."(Lives, xvi). One can say that the voices and thoughts that *Melanctha* sounds result in the most unconventional and challenging of the three stories. Stein tells us that *Melanctha* overthrows the reign of the nineteenth century. The text replaces the Victorian virgin or upstanding matron, protagonist with, a twentieth century, sexually liberated woman who creates her own world view, and lives by her own rules and objectives. In Melanctha's world of mixed races, "other world language," and "different ways," other potential possibilities for exchange between females, and females and males undergo a process of re-negotiation. One might say that Melanctha mediates a devenir-femme for her lover Jeff Campbell but as Deleuze tells us even "the woman" enters a becoming woman. By entering this becoming, the woman "gets out of her past, future and history."[4]

A moralizing, conventional interpretation of *Melanctha* particularly at the time it was written,1909, would surely have reduced it to an a propos study of the degenerate, destructive behavior that goes on freely in the more uncivilized, licentious, American minority groups. However, Stein's text writes against such interpretations because it subverts all the moralizing schemas of traditional society. Melanctha "wanders" with Jane Harden and "wanders" with a variety of men that span a range of classes and colors. However she does so not because she seeks pleasure, financial security or comfort but because she experiences a primordial desire and drive for *understanding* and *wisdom.*

> "Melanctha Herbert had always been old in all her ways and she knew very early how to use her power as a woman, and yet Melanctha with all

> her inborn intense wisdom was really very ignorant of evil. Melanctha had not yet come to understand what they meant, the things she so often heard around her, and which were just beginning to stir strongly in her. . . . Melanctha began to know her power, the power she had so often felt stirring within her and which she now knew she could use to make her stronger. . . . Melanctha almost forgot to hate her father, in her strong interest in the power she now had within her. . . . Melanctha now really was beginning as a woman. She was ready, and she began to search in the streets and in dark corners to discover men and to learn their natures and their various ways of working. In these next years Melanctha learned many ways that lead to wisdom. She learned the ways, and dimly in the distance she saw wisdom. These years of learning led very straight to trouble for Melanctha, though in these years Melanctha never did or meant anything that was really wrong." (Lives, 66, 67)

The point of the lengthy citation above is to give the reader a sense of how many times over a short course of four paragraphs Stein associates Melanctha's particular behavior with the signifiers; power, learning, strength, wisdom, discovery. In terms of a moralizing Victorian context nothing makes sense in this "scene." In the former context Melanctha would be characterized as a "run around," "a fallen woman" or "a tramp," but in Stein's text nothing she does is "really wrong" and she has no innate sense of evil or guilt. Furthermore the narrator describes her as intelligent, graceful and sweet. None of the conventional adjectives typically assigned to untamed women such as, loose, weak, simple, or wayward, apply. In fact the text never characterizes Melanctha in pejorative terms. "Melanctha was now come to be about eighteen years old. She was a graceful, pale yellow, good looking, intelligent, attractive "negress," a little mysterious sometimes in her ways, and always good and pleasant, and always ready to do things for people." (Lives, 75).

The "minor" language of Stein's brand of "black" vernacular sounds strange but coupled with a subversion of sign and signifier it takes the reader even farther into an-other world. This world takes form as Stein achieves the goal of her life's work which Weiss tells us was to "destabilize meaning at its most fundamental level, to illustrate the arbitrary relationship between the sign and the signified."(Stein and Wright, 35). Melanctha "wanders" with many men but rather than being portrayed as an object of men's desires, as a fallen, weak, victim, Melanctha deliberately seeks out these exchanges as part of an experiential quest that leads to self-expansion, empowerment and wisdom. "Melanctha had a strong respect for any kind of successful power. It was this that always kept Melanctha nearer in her feeling toward her virile and unendurable black father, than she ever was in her feeling for her pale yellow, sweet-appearing mother. . . . In these young days, it was only men that for Melanctha held anything there was of knowledge and power. It was not from men however that Melanctha learned to really understand this power." (Lives, 66, 67). Melanctha senses power and knowledge in men, but a woman, Jane Harden, teaches her *to really understand this power.*

> "Jane loved Melanctha and found her always intelligent and brave and sweet and docile, and Jane meant to, and before the year was over she had taught Melanctha what it is that gives many people in the world their wisdom. Jane had many ways in which to do this teaching. She loved Melanctha hard and made Melanctha feel it very deeply. She would be with other people and with men and with Melanctha, and she would make Melanctha understand what everybody wanted, and what one did with power when one had it. . . . In every way she got it from Jane Harden. There was nothing good or bad in doing, feeling, thinking or in talking, that Jane spared her. Sometimes the lesson came almost too strong for Melanctha, but somehow she always managed to endure it and so slowly, but always with increasing strength and feeling. Melanctha began to really understand." (Lives, 74)

Even the space of friendship or sexuality produces an in-between. "She would be *with* other people and *with* men and *with* Melanctha." The fact that these women are involved in an intimate relationship again pushes the limits of Stein's revolutionary terrain. As the absence of the white law enables the devenir-noir a more extreme intensity, similarly, the absence of the male in the sexual relationship propels this devenir-femme towards its furthest extreme. Because the text erases the representative of power and law being the white male, there are no more fixed places, there are no preordained patterns to life, love and exchange. After Melanctha "graduates" from Jane's course and becomes stronger than Jane, she goes back out to the world to "teach."

> "Then slowly, between them, it began to be all different. Slowly now between them, it was Melanctha Herbert who was stronger. . . . Melanctha Herbert was ready now herself to do teaching. Melanctha could do anything now that she wanted. Melanctha knew now what everybody wanted. . . . And so Melanctha began once more to wander. It was all now for her very different. It was never rougher men now that she

talked to and she did not care much to know white men of the, for her very better classes, it was something realer that Melanctha Herbert wanted. . . . With these men she knew she could learn nothing. She wanted some one that could teach her very deeply and now at last she had found him, she really had it, before she had thought to look if in this man she would find it." (Lives, 74, 75)

Melanctha believes that Jefferson Campbell, the young doctor who attends to her dying mother will be this person to further teach her "very deeply." However Melanctha, an excellent student, remains in the teacher position due to the fact that she has acquired more "real wisdom" and experience in her *ways* and *wanderings* as opposed to Jeff who has believed and thought about "one right way of living regular."(Lives, 85).

"She asked him many questions and always listened very well to all he told her, and she always remembered everything that she had learned from all the others. . . . Melanctha did not feel the same as he did about being good and regular in life, and not having excitements all the time, which was the way Jefferson Campbell wanted that everybody should be, so that everybody would be wise and happy. Melanctha always had strong the sense for real experience. Melanctha Herbert did not think much of this way of coming to real wisdom."(Lives 80, 81)

Melanctha's ability for "real feeling" and her desire for "real wisdom and real experience" distinguishes her from Jeff who has trouble understanding the complexity of her thought and states, "I am a quiet, slow minded kind of fellow, and I am never sure I know just exactly what you mean by all that you are always saying to me. But I sure do like you very much Miss Melanctha and I am sure you got very good things in you all the time."(93). These two constantly struggle to reach a middle ground of understanding when they exchange expressions of their proper ideas and feelings because fundamentally their minds and hearts work differently. "He was silent, and this struggle lay there, strong, between them. It was a struggle, sure to be going on always between them. It was a struggle that was sure always to be going on between them, as their minds and hearts always were to have different ways of working."(Lives, 108). Nonetheless Jeff does come to realize "with Melanctha Herbert he could learn to really understand, then he was certain that he did not want to be a coward."(Lives, 91). What Melanctha has to teach Jeff is to *think* and *talk* less, in order to *deeply feel* and understand, to recognize the discrepancy between what he designates as the "regular way of living" that he wants for himself and everybody else and the meaning of his own actions and attractions.

> "'But how about Jane Harden, seems to me Jeff Campbell you find her to have something in her, and you go there very often, and you talk to her much more than you do to the nice girls that stay at home with their people, the kind you say you are always wanting. It don't seem to me Dr. Campbell, that what you say and what you do have much to do with each other. And about your being so good Dr. Campbell. . . . You don't care about going to church much yourself, and yet you always are saying you believe so much in things like that, for people. It seems to me, Dr. Campbell that you just want to have a good time just like all us others. . . . No, Dr. Campbell, it certainly does seem to me you don't know very well yourself, what you mean when you are talking.'" (Lives, 82)

Melanctha constantly asks him if he ever stops thinking or talking long enough to "feel," to know things beyond the scope of his programmed, *right* ways. She explains, "I certainly never do talk very much when I like anybody really, Jeff. You see, Jeff, it ain't much use to talk about what a woman is really feeling in her. You see all that, Jeff better, by and by, when you get to really feeling. You won't be so ready then always with your talking." Melanctha slowly starts to convert Jeff to her ways but this conversion entails uncertainty and internal strife for Jeff Campbell.

> "'I don't ever say you ain't always right, Melanctha,' said Jeff Campbell. 'Perhaps what I call my thinking ain't really so very understanding. I don't say, no never now any more that you ain't right, Melanctha, when you really say things to me. Perhaps I see it all to be very different when I come to really see what you mean by what you are always saying to me..' . . . These months had been an uncertain time for Jeff Campbell. . . . He now never thought about all this in real words anymore. He was letting it fight itself out in him. He was now never taking any part in this fighting that was always going on inside him." (Lives, 95).

Through Melanctha's mediation of *other* ways, and thought patterns, Jeff enters a realm of uncertainty and change where he begins to even stop thinking of "all this," all she mediates, in "real words." In this space Jeff comes to know a love that vacillates between those two kinds of loving he previously defined as the only kinds of love, one being *right* the other one being *wrong.* Now Jeff feels love as made of new things, or all different little pieces that come together, things that he thought were bad with those that are good, all the elements he had compartmentalized in distinct categories come to merge and fuse in one good, big feeling, through Melanctha's "teaching." Jeff feels true happiness at this new experience but fails to maintain it because he cannot relinquish his past thoughts and beliefs. Guilt linked to his past ideals, settles in to challenge and taint his new found love.

> "You see Melanctha, its like this way with me. I got a new feeling now, you been teaching to me, just like I told you once, just like a new religion to me, and I see perhaps what really loving is like, like really having everything together, new things, little pieces all different, like I always before been thinking was bad to be having, all go together like, to make one good big feeling. You see, Melanctha, its certainly like that that you make me been seeing, like I never know before there was all kinds of loving to come together to make one way really truly lovely. I see that now, sometimes, the way you certainly been teaching me, Melanctha, really, and then I love you those times . . . and then it comes over me sudden, perhaps I am wrong now, thinking all this way so lovely, and not thinking now anymore the old way I always before was always thinking, about what was the right way for me to live regular and all the colored people, and then I think, perhaps Melanctha you are really just a bad one . . . and then I always get so bad to you, Melanctha, and I can't help it with myself then." (Lives, 112)

Jeff suffers inner turmoil because he *remembers* instead of feeling the present. Intermittently he recalls the account of Melanctha's past told to him by Jane Harden. He recollects how he used to believe, and think, and know differently the "right ways," before he met Melanctha. "What was it that Melanctha felt then, that made Jeff remember all the feeling he had in him when Jane Harden told him how Melanctha had learned to be so very understanding? Jeff did not know what it was that had happened to him . . . What was it Melanctha was now doing to him? What was it he used to be thinking was the right way for him and all the colored people to be always trying to make it right, the way they should be always living?"(109). Jeff wrestles with accepting Melanctha's past as recounted by Jane Harden and mourns the loss of his former ideals that grounded him in a sense of stable, secure knowing.

> "Jeff felt a strong disgust inside him; not for Melanctha herself, to him, not for himself really, in him, not for what it was everybody wanted, in them . . . he only had disgust because he never could know really what it was really right to him to be always doing, in the things he had before

believed in, the things he before had believed in for himself and for all the colored people, the living regular, and the never wanting to be always having new things. . . . It was all so mixed up inside him. All he knew was he wanted very badly Melanctha should be there beside him, and he wanted very badly, too, always to throw her from him. What was it really that Melanctha wanted with him?"(Lives, 110)

Melanctha comes to represent a radical space that causes Jeff's existential crisis. How can he feel so right with a woman who represents all he ever believed was the "wrong way to be living"? Melanctha and all she teaches him incites confusion because he can not "know" her and her motives due to the fact that "Melanctha was too many for him." (124). Jeff expresses his frustration at Melanctha's complex and paradoxical differences that she contains and diffuses that make it impossible for him to understand or know anything "real" about her.

"I certainly after all this time I know you, I certainly do know little, real about you. . . . Sometimes you seem like one kind of girl to me, and sometimes you are like a girl that is all different to me, and the two kinds of girls is certainly very different to each other, and I can't see any way they seem to have much to do, to be together in you. . . . And then certainly sometimes, Melanctha, you certainly is all a different creature, and sometimes then there comes out in you what is certainly a thing, like a real beauty. I certainly, Melanctha, never can tell just how it is that it comes so lovely. Seems to me when it comes it's got a real sweetness, that is more wonderful than a pure flower, and a gentleness, that is more tender than the sunshine, and a kindness, that makes one feel like summer, and then a way to know, that makes everything all over, and all that, and it does certainly seem to be real for the little while it's lasting, for the little while that I can surely see it, and it gives me to feel like I certainly had got real religion. . . . I certainly did think once, Melanctha, I knew something about all kinds of women. I certainly know now really, how I don't know anything sure at all about you." (Lives 97, 98)

After a certain point, as there seems to be nothing "sure that can be known" about Melanctha the woman, the same seems to follow for Stein's text.[5] It appears that in the strangeness of its language, relations, associations nothing can be known because the movements of these elements perpetuate flux and flow that can't be tied down to comprehensive formulas. The strangest quality perhaps is that rather than the characters "moving" or acting, the only textual elements that move, seem to be the words that circle and sound refrains of thoughts that vacillate to and fro, for the most part between Melanctha and Jeff Campbell. These words that shift, repeat, diverge, intermingle and separate, mediate the states, the multiplicities, through which to feel the lives of these characters. In fact Melanctha and Jeff's tension is textual rather than sexual. They verbally engage, couple, separate, and transpose their positions out of this process of exchange, in a lieu of vacillation.

What results from what may be interpreted as an "existential" debate, is that Melanctha teaches Jeff but he understands her lessons too late to live them with her. It appears that the text affirms Melanctha's views on the primacy of feeling and *real* experience that breeds the most intense and profound sense of life and a sort of "anti-memory, remembering." Once Melanctha realizes that Jeff will never understand, feel or *remember* as she does because he can't "forget the past," she realizes her position of strength again and has no option but to leave him.

> "You remember right, because you don't remember nothing till you get home with your thinking everything all over, but I certainly don't think much ever of that kind of way of remembering right, Jeff Campbell. I certainly do call it remembering right Jeff Campbell to remember right just when it happens to you . . . like that day in the summer when you threw me off just because you got one of those fits of your remembering. No, Jeff Campbell, its real feeling every moment when its needed, that certainly does seem to me like real remembering. . . . I certainly don't think much of the way you always do it, always never knowing what it is you are ever really wanting and everybody always got to suffer. No Jeff, I don't certainly think there is much doubting which is better and the stronger with us two, Jeff Campbell." (Lives, 128, 129,130)

The issue with "remembering" is an important one in Stein's text as indicated in the above citation: Stein repeats the signifier seven times in the short span of four sentences. When Jeff "remembers" the disparaging remarks about Melanctha that Jane Harden relayed, he reacts to Melanctha negatively without warrant and "throws her off." She understands how he "remembers" and communicates that it, makes him think poorly of her in the present even if in the present she is true, loving and loyal to him exclusively. According to Melanctha the best way to "remember" is paradoxically to *feel* the present moment. The present *feeling* if positive and "real," being a culmination of all former moments that lead to it needs no justification from references to the past. However Jeff who has a hard time "feeling deeply" because he "thinks too much," can not learn Melanctha's method of "remembering," therefore he fails to grasp or appreciate Melanctha in the moment. Jeff only realizes or feels his love for Melanctha deeply after she has gone, when he can *remember* her love. "It was now, when he had learned to really love Melanctha, she did not need any more to have him."(133). This important theme of "remembering," feeling, and combining diverse and different pieces to make a lovely new whole metaphorically refers to the mechanisms of Stein's textual composition that she describes as, writing a "continuous present." This *continuous present* can also be read as one of the primary qualities that allows for the devenir-femme and the devenir-noir, because the entry into a devenir-noir or a devenir-femme as Deleuze tells us entails a memory lapse. The devenir entails an experience out of time, divorced from past, future or history. "The line-system (or block) of becoming opposes itself to that of the point-system, that of memory. Becoming, is the movement through which the line liberates itself from the point, and renders the points indiscernible: rhizome, the opposite of arborescence, disengages from it. Becoming is an anti-memory. . . . The remembrance always functions as reterritorialization."(Mille Plateaux, 360). The devenir (becoming), like Melanctha's ways and wanderings trace out new orientations, and directions, entries in and out of the system beyond memory, beyond the history of the "race," the "man," the "woman" or the "sexes"; *becomings* that operate silently, imperceptibly. As Deleuze indicates even the woman although part of the minority must enter "les devenirs–femme" to escape and undermine the socially encoded Molar Order "block-woman's" past and future. (Dialogues, 8)

The introduction to Stein's *Three Lives* writes that Stein's virtually static or non-existent plot structure was not due to poor writing but was the result of Stein's deliberate refusal to write a cause and effect narrative. (Lives, xiii). As Stein herself emphasized, she intended to compose her piece after a fugue by Bach, or a portrait painting by Cézanne or Picasso. The importance of Cézanne's model of portrait painting was that every color played an instrumental part in creating the final whole: The pieces and parts actualize a whole that takes on numerous "non-whole" forms. Instead of a palette of colors and hues, Stein combines her pieces with ear "tones." "In one of her later essays Stein relays that the key to understanding her books lies in recognizing that she wrote "by ear" rather than "by eye": Instead of imagining pictures of what she describes in words one must concentrate on hearing the words she puts down on paper, to achieve in prose what she called a "continuous present." (Lives, viii, xiii).

The characters of Melanctha that move through words, glimpses of thought, feeling, a curious combination of sounds and sentences, without "memory" or action, resonate with Deleuze's description of characters from a

Thomas Hardy novel: His characters are collections, packages, blocks of various sensations rather than "people" or "subjects." These characters confront an unpredictable destiny in an "empiricist experimental world." These individuations or packs of sensations, run along and distend "lines of earthly deterritorialization": *they betray the fixed powers that want to hold us back, the established powers of the earth.* (Dialogues, 52).

Melanctha particularly moves as an agencement, as a non-material sign that inspires transformations, flux and flow. Melanctha described by Jeff as "too many" contains an abundance of differences due to the fact that she conjugates with a plethora of "ANDS." She moves between Gertrude Stein's two other *Lives,* she moves in and out of conjunctive relations, gets lost, and passes away but nonetheless her kind of memory of "deep feeling" of a continuous present remains "present," and active in the pages of the text. *Melanctha,* a sign, a medium of "devenir-noir," "devenir-femme" vacillates in, out and beyond a system of social, legal, moral and historical codes. This allows her to gain wisdom and power at a price: However, one can say that this is not without return. This return assures that those that enter Melanctha's devenirs have the opportunity to also gain some *other* world wisdom, become invited and inspired to see, hear, and think beyond the status quo markers that keep women and blacks in a certain time, place and historically predictable scenario. Through her gender-, code-, race- breaking discursive movements and language, Stein seduces the word and those taken in the words to venture beyond the major language's boundaries, laws, terrain, and worldview. But as the narrator tells us Melanctha has "break neck courage," and the life she leads indeed entails risk, danger and discomfort.

In fact Melanctha's life and story do not end well. Melanctha dies of "consumption" after a period of feeling lost, blue and suicidal. In a conventional, moral seeking reading, one might judge the end of Melanctha's story as one that finally sets things right and reasonable. Moralist readers might exclaim, "Finally, some poetic justice!" Melanctha dies not long after her friend Rose Johnson shuns her because Melanctha fails to behave in a "decent" manner. Paradoxically, the narration describes Rose, who has "strong the sense of proper and decent conduct," and gets "regularly really married," as "simple, sullen, selfish, unmoral, promiscuous, shiftless." Again, nothing makes "conventional sense."[6] In addition when Rose has a baby, it is Melanctha who tends to them with a "patient, submissive, soothing and untiring attitude." The text in fact in its opening and closing lines, in its structure of refrain, describes Melanctha as the representation of the nurturing, life sustaining force. For when Melanctha is absent for a few days, Rose's healthy baby dies because of the biological mother's selfish, neglectful nature. The narrator asks:

"The child though it was healthy after it was born, did not live long. Rose Johnson was careless and negligent and selfish, and when Melanctha had to leave for a few days, the baby died . . . Why did the subtle, intelligent, attractive, half white girl Melanctha Herbert love and do for and demean herself in service to this course, decent, ordinary, black childish Rose, and why was this unmoral, promiscuous, shiftless Rose married, and that's not so common either, to a good man of the Negroes, while Melanctha with her white blood and attraction and her desire for a right position had not yet been really married." (Lives, 60)

Melanctha as female sign, as bi-racial (white/black) sign, as text, in her/its complexity may in the end be read as a sort of tragic heroine, or heroic aesthetic act that moves as and *lives* as the sign, as the process of Stein's creative, experimental, revolutionary artistic temperament and philosophy. In *The Autobiography of Alice B. Toklas,* Stein states that the purpose of her refrains of strange sentences, dialogues and narrative descriptions is that they get under the reader's skin. "My sentences get under their skin, only they do not know that they do." One hundred and seventy pages later she repeats, "They do quote me, that means that my words and my sentences get under their skin."(70). This getting *under the skin* aspect of Stein's work evokes Deleuze's comment that the rhizome does not use the "subconscious" or interpret it but produces a subconscious with new, different desires and enunciations.

"For the enunciations as well as for desires, the question is never to reduce the unconscious, to interpret or make it signify according to the tree. The question is to produce the unconscious, and with it, new enunciations, other desires: the rhizome is this production of the unconscious itself."

« Pour les énoncés comme pour les désirs, la question n'est jamais de réduire l'inconscient, de l'interpréter ni de le faire signifier suivant un arbre. La question, c'est de produire de l'inconscient, et, avec lui, de nouveaux énoncés, d'autres désirs : le rhizome est cette production d'inconscient même. » (Mille Plateaux, 27)

Stein's own refusal of the "Freudian subconscious" and her insistence on writing refrains of rhythm and tones, that results in what she likens to a musical composition seeps below the surface to generate an other-consciousness whose irreducible residues generate revolutionary desires and meanings. "I don't hear a language. I hear tones of voice and rhythms. . . . By refusal of the use of the subconscious Gertrude Stein achieves a symmetry which has a close analogy to the symmetry of the musical fugue of Bach." (Autobiography, 70, 80)[7]

IN THE WAKE OF BETRAYAL: *DARK LAUGHTER, CANE,* AND *THE SOUND AND THE FURY*

Although Stein had to pay to have *Three Lives* published because of a poor reception from publishers who demanded she correct her English, the text eventually gained notoriety in its reprinted edition in 1933. This publication presented it as a "new classic." (Lives, xix). Ann Charters writes about the influence Stein's *Melanctha* had on the younger generation of moderns. "It was Stein's stubborn insistence on using her own language and her sympathetic view of the hardships of immigrant and black life in America that so stunned readers like Sherwood Anderson and Richard Wright that they championed her writing in the early decades of this century."(xviii). In his publication, "I Wish I'd Written That," Richard Wright describes the story as, "The first realistic treatment of Negro life I'd seen when I was trying to learn how to write. . . . This story made me see and accept for the first time in my life the speech patterns of Negroes, speech that fell all around me unheard." The line of influence that runs from Stein (the homosexual, Jewish, female, ex-patriot American) to Wright, affirms Deleuze's description of devenirs that inadvertently challenge essentialists' claims. Again this concerns the fact that men can activate a devenir-femme, and white writers a "devenir-noir." (Dialogues, 55). Surely Wright had read fiction written by African American writers before reading Stein, however it is Stein who "writes by ear" that renders the most "realistic treatment of Negro life."

Despite Stein's aesthetic achievements she did not live completely free of racial prejudice. Her own racial bias comes to the fore in the *Autobiography* when she groups all Negroes together under the sign of a 'narrow,' African cultural heritage. "Gertrude Stein concluded that Negroes were not suffering from persecution, they were suffering from nothingness. She always contends that the African is not primitive, he has a very ancient but a very narrow culture and there it remains. Consequently nothing does or can happen." (Autobiography, 238). Nonetheless as already stated, the devenir may be enacted by anyone even it that person carries traces of prejudicial ideas, as long as the writer "loses face," enters his/her own becoming-everyone, as each one, becoming- minoritaire. As Deleuze points out, Lawrence and Miller live as *phallocrats* however their writings still managed to lure them and others into an *irresistible "devenir-femme*." In addition, one can not assume that racist ideologies only infected and affected the psyches of white Americans. For really no one raised in America thinks or lives "racially free." American blacks and other minorities particularly those raised before the termination of the Jim Crow regime and racial propaganda campaigns also harbor traces of racism. The African American artist Wallace Thurman's novel *The Blacker the Berry* examines this issue as it takes the phenomenon of inner race racism as its central theme. The main character Emma Lou's own light skinned mother and grandparents who consider themselves "aristocratic Negroes" curse Emma's blackness and make her feel inferior.

> "Despite the rancor of her mother and the whispering of her mother's friends, Jane hadn't really found anything to regret in her choice of a husband until Emma Lou had been born. . . . She was abysmally stunned by the color of her child, for she had been certain that since she herself was so fair that her child could not possibly be as dark as its father. . . . But she hadn't reckoned with nature's perversity, nor had she taken under consideration the inescapable fact that some of her ancestors too had been black, and that some of their color chromosomes were still imbedded within her." (Blacker, 14)

The next section of the chapter further explores this theme of activating a "devenir-minoritaire" as a writer loses face, as a writer "betrays." Through a comparative view that clarifies the distinction between a writer that cheats and a writer that betrays, particular narrative and stylistic techniques that write against racism, sexism, -isms and -ists in general, will be considered as they run their *lines of flight* against the grain of "binary" codes and definitions written into the major language to constitute its *Major System.* On the other hand, one will also try to see what constitutes "cheating" and how the writer that cheats only reiterates and reinforces those codes established by the Major Order that rely on essentialist ideologies of authenticity, racial difference, superiority and inferiority. The juxtaposition of texts that betray with those that cheat also demonstrates that texts categorized as "modern" do not automatically share the qualities specific to those Deleuze distinguishes as "minor" or "rhizomatic." This section will make it clear that "minor" has less to do with chronological separations between one historical period and another and more to do with the way writing works against, inhabits the Major System to break it down from the inside. Many "modern" texts entertain provocative themes and language; to break with the legacy of Victorian 19th Century bourgeois literature, nonetheless if the writing fails to expose *minor lives,* sound "minor cords," or extend *lignes de fuites,* it remains "major book" or "livre-racine."

As already cited, Sherwood Anderson found Gertrude Stein's *Melanctha* a source of inspiration. In 1919 in *The New Republic* Anderson writes, an article entitled, "Four American Impressions: Gertrude Stein, Paul Rosenfeld, Ring Lardner, Sinclair Lewis."

> "I went to visit an American woman, Miss Gertrude Stein, in her own large room in the house at 27 rue de Fleurus in Paris. . . . She is laying word against word, relaying sound to sound, feeling for the taste, the smell, the rhythm of the individual word. She is attempting to do something for the writers of our English speech that may be better understood after a time, and she is not in a hurry. . . . She is making new, strange and to my ears sweet combinations of words. . . . I have a kind of undying faith that what she is up to in her kitchen in Paris is of more importance to writers of English than the work of our more easily understood and more widely accepted word artists." (Anderson Reader, 429, 430)

The influence of Stein's *Melanctha* coupled with his enthusiasm for Jean Toomer's *Cane,* another text again with an almost, all "colored" cast, prompts Anderson towards his own attempt to write a novel about American *black life* that he entitles, *Dark Laughter.* Darwin Turner writes of the extent to which *Cane* had an impact on Anderson's work as he cites a letter that Sherwood Anderson wrote Jean Toomer. "I wanted so much to find and express something clear and beautiful I felt coming up out of your race but in the end gave up. . . . And then McClure handed me the few things of yours I saw and there was the thing I had dreamed of beginning."(Cane, xx). Anderson writes *Dark Laughter* while in New Orleans the town which also provides the geographical backdrop of the novel. At the same time William Faulkner begins to frequent Anderson and his literary circle there. "In 1924 he went to New Orleans where for the first time he met and mingled with literary people including Sherwood Anderson, who encouraged Faulkner to develop his own style, to concentrate on prose, and to use his region for material."(NA, 2031). It cannot be questioned that Anderson who recommends that Faulkner write out of his *region,* the Deep South of Mississippi, simultaneously shares his enthusiasm for *Cane* in its innovative approach to, and use of "black" material. In fact, in Sherwood Anderson's sketch *A Meeting South* (1924*)* a mid-westerner (Anderson) displaced in New Orleans becomes friends with "David" (William Faulkner), the former speaks of the potential of writing "nigger stories." "A great many northern men and women come down our way and when they go back North write about the

South. The trick is to write nigger stories. The North likes them. They are so amusing." (Anderson Reader, 400). These lines are not in quotations as are other lines when David and the "I" discuss, so the reader must assume that the narrator or the "I' articulates them. These words accurately describe the kind of story Anderson writes in the form of *Dark Laughter,* being a "nigger story."

Despite the fact that Anderson initially plays the role of literary mentor and advisor to both Toomer and Faulkner, the latter two writers' texts "betray" the *major order,* operate as truly "minor" texts, whereas Anderson's novel typifies that of the "professional writer." It operates as a "pure redundancy at the service of the established powers." (Dialogues, 55). For as Deleuze articulates the writing of the traitor always allows for les devenirs-minoritaires (*minoritarian*-becomings), whereas that of the cheater takes hold of and reasserts the validity of fixed properties. The cheater who fails to "lose face," imitates rather than creates and reterritorializes all his lines in terms of the pre-existing positions, hierarchies, formulas that already exist. Many writers dream of being traitors and think themselves such. "However despite this, they are only little cheaters."(Dialogues, 56)

Deleuze continues on to specify that the traitor's writing which "meets up with minorities" does not mean, that the traitor writes *for* or *about* the minority; to approach it as an object of study in order to summarize or freeze frame the minority in terms of origin, essence or socially pre-assigned, static traits or characteristics. It means rather, to meet and move towards a conjugal *deterritorialization* where the minority and the writer each in turn, pushes the other, towards an *other* "becoming." A minority never exists *all done:* The *lines of flight* constitute it while they also enable it to *advance and attack.* (Dialogues 56).

These prerequisite conditions to writing "minority becomings" to write as *traitor,* perhaps explain why Toomer and Faulkner succeed in writing as "non-professionals" whereas Anderson writes as the *cheater.* For Anderson who desires to express something out of an exploitation of "black material," reaffirms the lines that separate white and black into opposing racial categories, and retraces lines of the "established powers or order." He suggests that "race" exists as a certain possessed and possessing state. "I wanted so much to find and express something . . . out of *your* race." Toomer and Faulkner have more fluid notions of racial designations. As Deleuze points out, Faulkner himself speaks of his decision to "become black." "As Faulkner said, he had no other choice than to *become-Negro,* in order not to find himself fascist."(Mille Plateaux, 358). Jean Toomer, who publishers, critics, and fellow writers labeled "colored," although he "passed for white," resisted being identified with either race. Toomer spoke of himself as belonging to the "new race," the "American race," a mixture of a variety of nationalities and blood strains. In *Color and Culture* Ross Posnock documents Toomer's response to Anderson's limited and limiting essentialist notions of race and origin. "Sherwood limits me to Negro. As an approach Negro is good. But try to tie me to one of my parts is surely to lose me. My own letters have taken Negro as a point, and from there have circled out. Sherwood . . . ignores the circles."(Color, 32). Toomer "strives" for a *fusion* that may be read in light of Deleuze's term that opposes "conjunction" to "imitation"; conjunction signifies a deliberate interior "*spiritually* integrated" reception and distillation of variant particles in flux. If Toomer and Faulkner achieve "black-becomings," "les devenirs-noirs," in their writings it is certainly not because of skin tone, blood lines, or a belief that they are exclusively members of the black minority. Their ability to write as "traitors" rather is due to a style that mirrors the freedom of their psyches that manifests a belief in the possibility of shifting the boundaries and categories legislated by the established order through its customs, codes, and language.[8] Deleuze writes in *Kafka pour une littérature mineure,* that the writer, the anti-"homme-écrivain," the "experimenter," creates a multiplicity of devenirs through emphasizing variations in voice and sound: Or, through a style that releases these. (Kafka 15). Such a *style,* exemplified in the texts of Stein and Du Bois, releases the "minor" forms of language to "*fait langue,*" to make language; its inherent qualities disrupt the solidity and credibility of constants within the system of the "*public*" or "major language."

> "Proust said: "The chefs d'oeuvre are written in a sort of foreign language." To be a foreigner, but in one's own language, and not simply as someone who speaks a language other than his own. To be bilingual, multilingual, but in one and the same language. It is here that the style "fait langue"(makes language). It is here that the language becomes intensive, a pure continuum of values and intensities. It is here that language becomes secret, while it has nothing to hide. The continual variation has only ascetic lines. . . . These secret languages are chromatic, close to a musical notation. A secret language not only has a number and a hidden code . . . and forms a "sous-system," (an under-system); it puts the system of variables of the public language in a state of variation." (Mille Plateaux, 123,124)

Deleuze describes the writing that "makes language," in the same terms he employs to speak of music, which for him represents the ultimate "rhizomatic" medium; the medium that most dynamically proliferates variations to disrupt the reference points, the foundations of the "major order." "While putting all components in continual variation, music itself becomes a sub-linear system, a rhizome instead of a tree, and serves a cosmic virtual continuum, wherein even the gaps, silences, ruptures, cuts play a part. . . . There is only music which is art as cosmos, and traces virtual lines of infinite variation." (Mille Plateaux, 121). Deleuze also writes that the more writing expresses multiple forms which favor dynamic differences rather than static linguistic markers, grammatical, syntactic, lexical rules, it not only comes closer to "musical notation" but to music itself. (Mille Plateaux, 132). Faulkner and Toomer's texts operate as music so defined, where sound and voices are emphasized, and gaps, silences, ruptures and cuts play a part to trace and chase virtual lines of infinite variations. Their writing styles value, and release "la petite musique," which also resonates through the works of writers such as, Proust, Kafka, Du Bois, Stein, Celine, Beckett, and Nietzsche.

> "He was speaking of "la petite musique" (Celine). Kafka aussi, it is *la petite musique,* another, but always deterritorialized sounds. . . . These are the true *auteurs mineurs.* A way out for language, for music, for writing. . . . To make use of *polylinguisme* in one's own language, make of this a minor or intensive usage, oppose the oppressed character of this language with its oppressive character, find the points of non-culture and under-development the third world linguistic zones through which a language escapes itself" (Kafka 49, 50)

This paper turns to the texts of Toomer, and Faulkner to provide examples of writing styles that value sound, voice, variations and transmit the minor, intensive, "third world" linguistic zones where language escapes itself and its oppressive, oppressing limits in order to write against, the establishment, its history and memory. These writings as Kafka's "petite musique," know only lines that displace and efface points to result in a proliferation of irreducible meanings.

Literary critics refer to both Toomer's and Faulkner's writings as innovative and groundbreaking in that they generate a novel mode of fiction writing. Darwin Turner who writes the introduction to *Cane* points out that although critics generally categorize the work as a novel, it resists being labeled as any one genre as it comprises poetry, verses of songs, prose sketches and a drama piece. Despite a certain thematic consistency in the text each section stands complete and autonomous on its own. There is no carry over of characters from one sketch or section to the next. In fact

Toomer added the entire second section of *Cane* only because publishers complained that the work was too short. Moreover this added section not only writes of new and different characters, the geographical location shifts from rural Georgia to the urban Negro neighborhoods of Chicago and Washington D.C.. Turner writes, "To study Toomer's work only in relationship to the New Negro Renaissance, however, is to misjudge its value in the total culture of the twenties. Like other writers of the twenties, Toomer questioned the harmonies and values of his society. *Cane* is no conventional world of black primitives or exotics. It is a montage. . . . Stylistically too, Toomer rode in the vanguard of his generation. . . . No matter what name is given to the book's form, Jean Toomer did not conceive *Cane* as a novel." (Cane, xxi). In a like manner, the *Norton Anthology* points to Faulkner's ability to "invent" voices, his mastery of ventriloquism, and his unique techniques for representing time, states of consciousness and memory; literary approaches that extended the boundaries of fiction writing.

> "In each of the novels William Faulkner published between 1929 and 1936 it seemed as though fiction were being reinvented. He wrote about childhood, families, sex, race, obsessions, time, the past, his native South, and the modern world. He invented voices for characters ranging from sages to children, criminals, the insane, even the dead-sometimes all within one book. He developed, beyond this ventriloquism, his own unmistakable narrative voice, urgent, intense, highly rhetorical. He experimented with narrative chronology and with techniques for representing mind and memory." (NA, 20, 30)

Another source on Faulkner also emphasizes his ability to "develop languages." "Faulkner developed a language (or languages) which offered resistance to the South's language; for that reason he was more honored beyond its borders than within." Faulkner's talent to develop a plurality of languages that destabilize the *standardized, white* supremacist language of the South, his propensity to create different voices or to *become* other voices or *others* fictionally, along with writing a style that disrupts standard notions of time, and whose narrative proceeds through discontinuity are also qualities and techniques common to Toomer's *Cane.*[9] The characteristics that align these two works and enable them to operate as "writings of betrayal," distinguish them from less-modern works or those of "little cheaters."

Both *Cane* and *The Sound and the Fury,* structurally and "narratively" speaking, move forward through a series of breaks and discontinuities in both temporal and geographical zones and in narrative style or voice. Parts of

Cane completely efface the narrator function for instance when the text breaks into poetry, song and drama. When there is a narrative voice it shifts or changes from an impersonal to personal narrative position depending on the sketch. Faulkner's novel splits into four sections that are each narrated by a different voice, the first three by characters and the final section by an unidentifiable voice. The narrative styles of Benjy and Quentin, are especially discursive due to the fact that Benjy is both literally and linguistically castrated, a subject who only bellows and cries to the exterior world, and Quentin delivers his account on the same day that he prepares for his suicide. These texts vacillate back and forth between present and past without warning or logical linkage between the movement across time zones. The narrative carries the reader along within the flow of these characters' inner streams of consciousness. Perhaps it goes without saying that the three Compson brothers who each narrate a different section remember their family history differently. Due to these characteristics, both works resist being reduced to one theme or meaning while they disqualify the credibility of the narrator function, and the reliability or infallibility of memory and history.

In opposition to *Cane* and *The Sound in the Fury,* Anderson's novel proceeds as a story with a beginning, middle and end that progresses through a succession of chronologically ordered events. The novel's climax comes at the end when the heroine of the story, Aline Grey leaves her husband to be with her gardener/lover Bruce. The reader follows along as a passive participant as the action unfolds due to the narrator's omniscient position. This kind of writing resembles Deleuze's description of "arbre-racine," or the "cheater's" writing that reinforces the dichotomies, and hierarchical structure of the established order. (Dialogues, 33). Due to this orderly structure and the narrator's authoritative, consistent and continuous position within the novel the reader accepts the validity of the narrator's insights into each character's inner life. This "readerly writing" (Barthes)[10], that grants the narrator authority also lends more seriousness to the novel's description of blacks: "The niggers were something for Bruce to look at, think about." Turner accurately terms Anderson's portrayals of blacks as pejorative: Indeed Anderson does present black Americans as objects as he reduces them to "primitive," one-dimensional figures.

In his introduction, Turner not only refers to the influence *Cane* had on Sherwood Anderson's move to join in the trend of exploiting black people as subjects and themes in modern fiction. He also mentions Faulkner's *The Sound in the Fury* as another example of this trend among white artists, however he does not describe Faulkner's portrayal of Blacks as "primitive," "innocent" and "laughing," in other words, as the stereotypical early twentieth century cultural rendition of black subjectivity. Instead he refers to Faulkner's presentation of Blacks as "solid" and "stable" in contrast to the Whites who are instead "neurotically over-sensitive." This reading agrees that Faulkner doesn't present a one-dimensional stereotypical caricature of Blacks but insists that these "stable" blacks are not stable in the sense of static, or represent a generic fixed type; that the one stands for the many, the group. On the contrary, in the variations of "blackness" Faulkner presents, he destabilizes any notion of there being one standard definition of Black or White identity and simultaneously alters the lines that mark out and separate these two groups as binary opposites.

In *Cane* and *The Sound and the Fury,* the succession of different voices, an absence of one main voice employed as a centralizing, organizing force, allows for a dispersal and value given not only to third-world, minor-language variations but also gives acknowledgement to the heterogeneous groups and their subjects who put these variations into play. "Only the minor is grand and revolutionary. Hatred for all the literature of the masters. Kafka's fascination for servants and employees (the same thing in Proust for servants, for their language). . . . To be a foreigner in one's own language. . . . In sum, the German of Prague is a deterritorialized language, proper to strange minor usages (cf., in another context today, that which Blacks can do with American)." (Kafka, 48, 30). The voices of servants, of minor or socially muted subjects are unarguably given sound in both *Cane* and *The Sound and the Fury,* whereas in Anderson's *Dark Laughter* the narrator's pen never registers direct speech or dialogue of the black subjects/objects. Out of all the black characters who fill his pages he only points to two black subjects as distinct individuals in his "black folk," homogeneous grouping. Nonetheless the text presents these black women who serve in Aline's household as mute signs that represent the writer's views on the "black group." Through their opaque *objectivity* Anderson tells us what black people *essentially* "are" and "have." "There were two servants in the house but they were both negresses. Negro women have an instinctive understanding."(DL, 233). "No one else about the house but two negro women. Negro women have no moral sense. They will do anything." (DL, 237). "The two Negro women were also going. Presently they went down along the path to the gate. For them it was a gala occasion . . . Nigger women prancing for nigger men. "Come on baby!"(DL, 262). "They are like children looking at you with their strangely soft innocent eyes. White eyes, white teeth in a brown face -laughter."(DL, 266). The only sounds black people make in this text which is supposed to "express something beautiful coming from *Toomer's* race," are laughs and voices in bits of songs. These songs however, along with the laughter, only serve to "ghettoize" black expression, reduce blackness to a sign that signifies; happy, primitive, exotic, "darky." For as Gilles Deleuze emphasizes minor languages do not "imitate" one form or aspect of a minority's "language." This would only reinforce major order stereotypes. One must rather disseminate the plurality of communicative, signifying modes at work within the minor order.

> "We must distinguish: the majority as a homogeneous constant system, the *minoritaire* as a potential, creative becoming. . . . This is the continuous variation that constitutes the devenir-minoritaire of everyone. . . . This is certainly not in using a minor language as a dialect, *making* regionalism or ghetto, that one becomes revolutionary; It is by using many minority elements, connecting, conjugating them that one invents a specific, autonomous, unexpected becoming." (Mille Plateaux, 134–5).

Anderson like Toomer intersperses bits of "Negro song" into his text but the verses he chooses simply reinforce the narrator's rendition of black people as laughing, singing, innocent children.

> "Soft voices laughing, laughing, "Oh, ma banjo dog, Oh, ho, ma banjo dog." . . . Ah, my baby! Ah, my baby!" Sounds caught in black throats. Notes split into quarter notes. The word, a meaning, of no importance. Perhaps words were always unimportant. These were strange words about a "banjo dog." What was a "banjo dog?" Ah, my banjo dog! Oh, oh! Oh my banjo dog!." . . . Negroes singing- "An" the Lord said . . . Hurry, hurry. Negroes singing had sometimes a way of getting at the ultimate truth of things. Two Negro women sang in the kitchen of the house. . . . About the Negro women it did not matter. They would think as their natures led them to think, feel as their natures led them to feel. You can't tell what a Negro woman thinks or feels." (DL, 73, 106, 266)

Although Anderson uses "black song" to speak of extra-verbal meanings that may be passed on by the voice regardless of what the words signify, a point also stressed in *Cane* and *The Sound and the Fury,* Anderson makes this point by contrasting black and white. Blacks sound a pre-linguistic sense of the primitive, careless state of being through songs that reflect a "natural" aversion to linguistic complexity. The Whites on the other hand, are over burdened with a constant state of reflection that hinders them from looking at life through the eyes of "innocent children." The narrator repetitively describes blacks with the following verbs: singing, laughing, shuffling, and dancing. But, there is no interiority, they move only as; bodies, colors, sounding mindlessly, free of any existential angst or complex consciousness. "Standing, laughing—coming by the back door—with shuffling feet, a laugh—a dance in the body."(DL, 74). In relation to this kind of fictional caricature one must call to mind Ralph Ellison's appropriate words. "Too often what is presented as the American Negro (a most complex example of western man) emerges an oversimplified clown, a beast or an angel. Seldom is he drawn as that sensitively focused process of opposites, of good and evil, of instinct and intellect, of passion and spirituality, which great literary art has projected as the image of man."(Shadow, 26).

The songs and sounds released within the pages of *Cane* and *The Sound and the Fury* transgress diametrically opposed racial categories and associated stereotypes based on binary oppositions. In *Cane,* the play, entitled, "Kabnis" disperses a plurality of sounds and silences that break apart the categories that separate people along racial lines, and de-legitimize groupings based on race, that lump all who contain "black blood" into one homogeneous grouping.

From the opening lines of the drama, Ralph Kabnis demonstrates an intense level of discomfort and fear in an environment that encompasses and vibrates with the "weird chill" of the local Negro folk songs of rural Georgia. In the opening scene, Ralph Kabnis tries to read himself to sleep, but must listen to the sounds that seep through his lodgings, that communicate the southern legacy of violence, racism, and slavery. "Ralph Kabnis, propped in his bed, tries to read. To read himself to sleep . . . cracks between the boards are black. These cracks are the lips the night winds use for whispering. Night winds in Georgia are vagrant poets, whispering. Kabnis, against his will, let his book slip down, and listens to them. The warm whiteness of his bed, the lamp-light, do not protect him from the weird chill of their song."

> White-man's land.
> Niggers, sing.
> Burn, bear black children
> Till poor rivers bring
> Rest, and sweet glory
> In camp Ground.
> (Cane, 81)

These songs communicate something beyond "whiteness," the "warm whiteness" of his bed and lamplight that should protect him as they hold and surround him. Throughout the piece, Kabnis struggles with a world foreign to him that disseminates such chilling strange meanings that stretch beyond major protection.

This play relays that "whiteness," its laws, definitions, protective mechanisms, are no more static and fixed than blackness. Furthermore, whiteness and blackness do not refer to two mutually exclusive states of being. Kabnis's fear born of his geographical and cultural displacement is compounded when he frequents Layman and Halsey, two middle class southern Blacks who introduce him to the white definitions of "Negro" in Georgia. Kabnis wants these men to reassure him that his status as an educated, northern black which makes him a "gentleman" will guarantee him safety from the race hate crimes of the South. "Kabnis: 'But they wouldn't touch a gentleman—fellows, men like us three here—.' Layman: 'Nigger's a nigger down this away, Professor. An only two dividins: good an bad. An even they aint permanent categories. They sometimes mixes um up when it comes t lynchin. I've seen um do it.'" (Cane, 87).

Not only Kabnis's composure but his speech patterns shift to resemble those of Layman and Halsey as he descends into the latter's basement and meets the old, unidentifiable ex-slave believed to be deaf, dumb and blind, he who dishevels Kabnis's emotional state with his presence and inarticulate musings. This scene juxtaposes Kabnis to Lewis, another educated, light skinned northern Negro who finds himself teaching in rural Georgia. Lewis is described as Kabnis's double, as a sort of alter ego. "Lewis enters. He is the queer fellow who has been referred to. A tall wiry copper-colored man, thirty perhaps. His mouth and eyes suggest purpose guided by adequate intelligence. He is what a stronger Kabnis might have been, and in an odd faint way resembles him." (Cane, 95).

Lewis the "stronger Kabnis" characterizes the old man as a mystical link to slave days and a blind, muted prophet that represents the silenced voices of all of those forsaken victims of slavery. "Slave boy who some Christian mistress taught to read the Bible. . . . Dead blind father of a muted fold who feel their way upward to a life that crushes or absorbs them. (Speak, Father!) Suppose your eyes could see, old man (The years hold hands. O Sing!) Suppose your lips. . . . Halsey does he never talk?" Kabnis who is repelled and irked by the old man's presence instead calls him, "Father of hell." Lewis continues to say to Kabnis, "The old man as symbol, flesh, and spirit of the past, what do you think he would say if he could see you?" Kabnis retorts, "An besides he ain't my past. My ancestors were Southern blue-bloods-." Lewis points out Kabnis's denial and fear of what this man of a muted world and experience represents as Lewis retorts . . ."and black." Kabnis hopelessly denies his denigrated status that the South imposes replying, "ain't much difference between blue and black." Lewis: "Enough to draw a denial from you. Can't hold them, can you? Master; slave. Soil; and the overarching heavens. Dusk; dawn. They fight and bastardize you."(Cane, 107).

The patriarchal figure of the group Halsey, calls upon Lewis as a kind of psychic sage to tell Kabnis what he "knows." Halsey: "Tell him Lewis, for godsake tell him. I've told him. But its somethin else he wants so bad I've heard him downstairs mumblin with th old man.." . . . Lewis: "Life has already told him more than he is capable of knowing. It has given him in excess of what he can receive. I have been offered. Stuff in his stomach curdled, and he vomited me." (Cane, 99). Lewis is the surplus, a radical supplement that Kabnis can not digest. The side that speaks truths that circulate beyond "knowing." Kabnis articulates his agitated state of being and response to this with his body. "Kabnis' face twitches. His body writhes."(Cane, 99). Kabnis: "You know a lot, you do." Kabnis must violently try to suppress the living reality of this ex-slave who represents the violent truth of a soil that "bastardizes" him as Lewis puts it. Whereas Lewis distinguishes and specifies Father John's identity, Kabnis who calls him "Father of hell," his psychic hell, speaks of him in terms of death, as he verbally attempts to annihilate him.

> Kabnis: "You sit there like a black hound spiked to an ivory pedestal. An all night long I heard you murmurin that devilish word. They thought I didn't hear y, but I did. Mumblin, feedin that ornery thing that's livin on my insides . . . youre dead already. Death . . . What are y throwin it in my face for? Whats it goin t get y? A good smashin in th mouth, that's what. . . . What do I care whether you can see or hear? You know what hell is cause you've been there. It's a feelin an its ragin in my soul in a way that'll pop out of me an run you through, an scorch y, an burn an rip your soul. Ha nigger soul. . . . Aint surprisin th white folks hate y so. . . . Oh, I'm drunk an just as good as dead. . . . Youre an old man, a dead fish man, an black at that. They've put y here t die, damn fool y are not t know it. Do y know how many feet youre under ground?. . . . Do y think youre out of slavery? Huh? Youre where they used t throw th worked-out, no-count slaves. On a damp clammy floor of a dark scum-hole." (Cane, 114).

Father John's presence evokes truth which vacillates in the point counter point sort of exchange that runs between the two light, educated black men who find themselves teaching in rural Georgia, and feel out of place among both the white and local black population. Kabnis claims his "blue-blood" heritage but denies his "slave-blood," because of the discrepancy between these two categories and all the other polar opposite categories that Lewis indicates can't be sanely "held" in one unified sense of identity. Either one is master or slave, "blue" or black, their incongruous pulling together makes the member who holds the two strains an outcast of the major order, a "bastard" within its language which sets up, confirms and legitimizes these rubrics as diametrically opposed.

In *Kabnis, Cane* emphasizes the reality of silent or incommunicable enunciations that extend from the inarticulate, utterings of the muted, blinded, man. Father John, the man who represents those who society, its history, laws and language buries alive, finally does speak and when he does he postulates on "white sin." The sin that created false fixed "histories" when it made the Bible, the American reference for order, law, racial difference, *lie.*

"Father John: Th sin whats fixed . . . (Hesitates).

Kabnis: Suppose youre talkin about that bastard race that's roamin round th country. It looks like sin, if that's what y mean. Give us something new an up t date."

Father John:—f tellin Jesus lies. O th sin th white folks 'mitted when they made th Bible lie."(Cane, 116).

The Sound and the Fury, similar to *Cane* releases the sounds of the *servi-teurs,* the employees, or in other words individuals who communicate from the margins, the kitchens, fields or basements, individuals who transmit minor, discursive forms of language and non-linguistic sounds. Dilsey the Compson's black servant's voice sounds itself through all four narrative sections despite their discrepancies. Unlike the songs that play throughout Anderson's *Black Laughter* that reflect the stereotype of the happy Negro, Dilsey's song transmits an effect closer to that of Du Bois's sorrow songs or the folk song that Kabnis hears in rural Georgia. "As she ground the sifter steadily above the bread board, she sang, to herself at first, something without particular tune or words, repetitive, mournful and plaintive, austere, as she ground a faint, steady snowing of flour onto the bread board."(Sound, 240). Dilsey's austere, mournful song is interrupted and challenged by the voices of her employers but it is not silenced. "Then in the kitchen Luster heard Mrs. Compson and Jason descending, and Jason's voice, and he rolled his eyes whitely with listening."(246). "Did you hear me?" Jason said. "I hears you," Dilsey said. 'All I been hearin, when you in de house. Ef hit ain't Quentin er yo maw, hit's Luster en Benjy. Whut you let him go on dat way fer, Miss Cahline?" (247). Dilsey confronts Jason who bullies the whole household, including his mother several times. "You's a cold man, Jason, if man you is. . . . I thank de Lawd I got mo heart dan dat, even ef hit is black."(187). Dilsey points to Jason's inhumane nature questioning his status as "man" and goes so far as to tell Jason what he *is.* Dilsey speaks in Black Southern dialect but the novel doesn't "ghettoize" this language because it and the languages the novel sounds resist being fixed in a contingent relationship based on color, as none of the many different language registers are spoken exclusively by any one race. Instead Dilsey's language transmits truths as Quentin's narrative suggests. "Dilsey said it was because Mother was too proud for him. They come into white people's lives like that in sudden sharp black trickles that isolate white facts for an instant in unarguable truth like under a microscope."(155). Dilsey, the true matriarch of the Compson household, raises all the children due to their own mother's fragile emotional and mental health, a fact she points out frequently. "And whur else do she belong? Dilsey says, "Who else qwine raise her cep me? Ain't I raised ev'y one of y'all?"(178). Dilsey's language, a form of "black English" like Kafka's Jewish Czech, does not reflect an impoverished variation of English, but rather refuses and usurps the restrictions imposed by constants. Instead of expressing a lack, it signals an ellipse, a moving paraphrase. One experiences a refusal of reference points, the dissolution of the standard form in favor of dynamic differences. "The more a language enters into this state, the closer it comes, not only to a musical notation, but to music itself."(Mille Plateaux, 131–132)[11].

It is Benjy's and Quentin's narrative sections that incessantly repeat Dilsey's words that insist on the value of non-linguistic knowing. In these chapters it is Dilsey that discloses the fact that Benjy who is deaf and dumb "smells things" among other unspoken truths. "He smell what you tell him when he want to." "Benjy knew it when Damuddy died. He cried. He smell hit. He smell hit." Don't have to listen nor talk." (Sound 85). Dilsey's words that repeat themselves countless times in the spontaneous free-flowing interior monologues of Quentin and Benjy underline the power and force of her words, her language, her song, that are digested on a subconscious level.

Faulkner not only speaks of ways to know and communicate beyond language to put language its laws, codes, definitions, into question, but also destabilizes an order that associates varying uses of language with exterior rather than interior states. For example in Quentin's narration two Northerners refer to him as speaking "black," or "Canadian?" "He don't talk like them. I've heard them talk. He talks like they do in minstrel shows." "You said he talks like a colored man." "Ain't you afraid he'll hit you."(Sound, 110). The story's last pages of the novel undermine the idea that linguistic registers are color or race specific. The visiting black preacher, similar to Quentin speaks a language that supposedly belongs to those on the *other* side of the "color line." The preacher's own skills of ventriloquism disrupt the notions that equate standardized language use with whiteness, and qualify white language as that which most effectively transmits "true" meanings or renders accurate versions of reality.

> "When the visitor rose to speak he sounded like a white man. His voice was level and cold. It sounded too big to have come from him and they listened at first through curiosity, as they would have to a monkey talking. . . . His arm lay yet across the desk, and he still held that pose while the voice died in sonorous echoes between the walls. It was as different as day and dark from his former tone, with a sad, timbrous quality like an alto horn, sinking into their hearts and speaking there again when it had ceased in fading and cumulate echoes. . . . He was like a worn small rock overwhelmed by the successive waves of his voice . . . and the congregation seemed to watch with its own eyes while the voice consumed him, until he was nothing and they were nothing and there was not even a voice but instead their hearts were speaking to one another in chanting measure beyond the need for words. . . . They did not mark just when his intonation, his pronunciation, became Negroid, they just sat swaying a little in their seats as the voice took them into itself." (Sound, 261, 262).

This chapter concludes to reaffirm that Gertrude Stein, Jean Toomer and William Faulkner's ability to write as traitors, to "betray their races," or rather the concept of race itself, results from their efforts to write innovatively, to produce a "writerly" text: To generate a textual lieu wherein the writer "loses face," activates a realm of a-signifying "non-sense" and "chromatic" sound, that seduces the reader into a variety of becoming-*other*s. Here one discovers a distinctly, different, aural atmosphere where minor-language variations penetrate under the surface and communicate *minoritarian* truths and realities. In this zone one encounters; the bi-racial, bi-sexual, wandering, mysterious, but sweet Melanctha, with her *subconscious-producing,* circular refrains; the blind, mute, ex-black slave in the basement, who finally speaks to clearly incriminate "white" sin based in white lies; the frightened, educated, mixed-race, northerner, Ralph Kabnis, whose own linguistic register shifts from "standardized English" to "black" vernacular, as he descends with resistance, to the sub-system-space, of "Father John"; Dilsey, the wise, black matriarch, her silent, austere songs and acute commentaries that relay penetrating meanings to differentiate her as the household's most credible and rational member; the deaf, dumb, linguistically and physically castrated Benjy, who "smells things" beyond real-realities; and the bigot, misogynist Jason and his static, reductive "major order" discourse: In juxtaposition to the "*minor*" characters, Jason and the order he represents validate the accuracy of Father John's words that hold "white sin" responsible for making language, law and the "bible lie." These writings of *betrayal* represent Deleuze's "superior" form of American "rhizomatic writing": writings that disseminate the sounds and voices that resonate and vibrate as "la petite musique" to release "minoritarian" particles and realities which destabilize and thwart the mechanisms of the "molar order." These "minor" writings, extend *lines of flight,* which escape fixed structures and generate becomings without past, future or memory: lines that resist and break the binary machine as they insert themselves between completely heterogeneous beings and elements to generate "unparalleled evolutions." (Dialogues, 34)

> "What remains is precisely their "originality," that is, a sound that each one renders (returns, gives), like a ritornello at the limit of language, but that it produces only when it takes to the open road (or to the open sea) with its body, when it leads its life without seeking salvation. . . . This is how Lawrence described the new messianism, or the democratic contribution of American literature: against the European morality of salvation and charity, a morality of life in which the soul is fulfilled only by taking to the road, with no other aim, open to all contacts . . . turning away from those that produce an overly authoritarian or groaning sound." (Bartleby, ou la formule, 112)

Chapter Four
Conclusion

> "Bergson answers the intellectualist conundrums by pointing back to our various finite sensational experiences and saying, 'Lo, even thus; even so are these other problems solved livingly.' When you have broken the reality into concepts you never can reconstruct it in its wholeness. . . . But place yourself at a bound, or *d'emblée,* as M. Bergson says, inside of the living, moving, active thickness of the real. . . . What really *exists* is not things made but things in the making. Once made, they are dead, and an infinite number of alternative conceptual decompositions can be used in defining them. But put yourself *in the making* by a stroke of intuitive sympathy with the thing and, the whole range of possible decompositions coming at once into your possession, you are no longer troubled with the question which of them is absolutely true. . . . Once adopt the movement of this life in any given instance and you know what Bergson calls the *devenir réel* by which the thing evolves and grows. Philosophy should seek this kind of living understanding of the movement of reality, not follow science in vainly patching together fragments of its dead results." (Pluralistic Universe, 261, 262, 263).

This quote leads the way out of this text, but "this way out," rather than serving as an exit that provides a conclusive closure, hopefully provides the reader with a sense of its directions and what may result from the movements of its "ways." In other words, by pulling the threads of this writing along rather than "wrapping them up," will better clarify the approach of this text and the meanings generated therein. Certain questions however should provide the reader with a sense of what this writer believes may best qualify the particularity of this study. What is the relationship between Deleuze and American literature and more broadly "*minor*" literature in general? How does Deleuze's philosophy and writing proceed in relation to what he names minor literary writings? How do Deleuze and his thought exonerate and extend *ways,* lines, signs, words and names that operate in synch with Virginia Woolf's prescription for reading/writing?

> "How Should One Read a Book. . . . Do not dictate to your author; try to become him. Be his fellow-worker and accomplice. If you hang back, and reserve and criticize at first, you are preventing yourself from getting the fullest possible value from what you read. But if you open your mind as widely as possible, then signs and hints of almost imperceptible fineness, from the twist and turn of the first sentences, will bring you into the presence of a human being unlike any other. Steep yourself in this, acquaint yourself with this, and soon you will find what the author is giving you, or attempting to give you. . . . Perhaps the quickest way to understand what a novelist is doing is not to read but to write; to make your own experiment with the dangers and difficulties of words." (Vwoolf, 259)

In other words, how and why does Deleuze write *with* literature, with other writers and philosophers, accept their gifts or offerings, follow their signs to enter and activate *other-becomings?* How does Deleuze manifest a reading that rather than judges, critiques or dictates, instead distends seminal lines to generate an expansive, experimental terrain? How does this writing manifest even if fleetingly, the qualities of "rhizomatic"-pragmatics and its ways, and what might be the value in such a process that reads to write *with* or conversely *writes with* to read? To respond to these questions, and better qualify the *Deleuzian way* and that of this text, this chapter will briefly revisit a few aspects of the thought, life, art, and style associated with, and activated by *Deleuze,* and his relationship to his *Master,* Jean-Paul Sartre. Finally this text hopes to present Deleuze's question: What qualifies an "act of resistance"? In what ever *minor* form this writing hopes to pay homage to Gilles Deleuze, to the *deleuzian way* as an "act of resistance" along with other such acts, generated by names such as; Wahl, Nietzsche, James, Stein, Du Bois, Faulkner, Toomer, Proust, and Woolf.

The first question in the above paragraph which concerns Deleuze's relationship to American literature and the rational behind this writing that applies various deleuzian *lines* and *theories* to the works of American minor authors, may best be responded to through the words of Rene Scherer. In the compilation of essays honoring Deleuze's life and work, *Tombeau de Gilles Deleuze,* Scherer contributes, « *Gilles Deleuze: l'écriture et la vie.* » In this essay Scherer cites Deleuze, "On écrit toujours pour donner la vie, pour libérer la vie ou elle est emprisonnée, pour tracer des lignes de fuite."(*On always writes to give life, to liberate life where it is trapped, for tracing out lines of flight*). He continues on to write that according to Deleuze, the most exemplary writing that gives and liberates life where it is trapped in order to trace lines of flight, escape, or disappearance, is that of the Anglo-American tradition. Scherer also emphasizes that this tradition inspires Deleuze's theories on writing and art.

> "In a certain way, each one of Deleuze's works can be considered as a theory of literature and writing, and, in particular, of Anglo-American literature . . . he entitled a chapter of *Dialogues:* "*Of the Superiority of Anglo-American Literature,*" and he considers this literature, in contrast to that of France, as that literature which was uniquely capable of liberating itself from *psychologism* and *moralism,* from the subject and the person, to give free reign to an auto-sufficient life, needing no other justification than itself. For Deleuze, literature is reference and source. . . . It is through a citation by Virginia Woolf or Charlotte Bronte that he illuminates his conception of the "*dispersion of the subject,*" of "*nomad singularities,*" of this dissemination of particles, molecules, which compose desire, the unconscious, the *molecular* machines."

> « D'une certaine façon, toute oeuvre de Deleuze peut-être considérée comme une théorie de la littérature, de l'écriture. Et, en particulier, de la littérature anglaise-américaine . . . il a titré un chapitre de *Dialogues:* « *De la supériorité de la littérature anglaise-américaine,* » et qu'il considère que cette littérature, relativement à la française, a été seule capable de se délivrer du psychologisme et moralisme, du sujet et de la personne, de donner libre cours à la vie auto-suffisante, n'ayant besoin d'autre justification qu'elle-même. La littérature est pour Deleuze, référence et source . . . C'est par une citation de Virginia Woolf ou de Charlotte Bronte qu'il éclaire sa conception de la « *dispersion du sujet,* » des « *singularités nomades,* » de cette dissémination des particules, molécules qui composent le désir, l'inconscient, les machines *moléculaires* » (Tombeau, 86, 87)

Although Scherer states that all of Deleuze's work can be considered as a *theory* of literature and writing, this *theory* must be seen as entirely unique as it proceeds in a manner contrary to "theory" as such. As Scherer underlines, what this relation to literature entails for Deleuze is a "source," a vehicle through which to express certain ideas which have a philosophical import. The second part of this relationship entails inspiration. This "source" not only inspires Deleuze to think about certain problems and work through them but provides a certain momentum or particular desire that produces the writing of them in a certain "way" that generates a plane of consistence or immanence. "What we tried to show, on the contrary, was how desire is a process, and that it is a *plan(e) of consistence,* a field of immanence. . . . Desire is not interior to a subject any more than it stretches towards an object: it is strictly immanent to a plan that doesn't pre-exist, a plan that needs to be constructed, where particles transmit themselves, and fluxes conjugate." (Dialogues, 108). Contrary to *theorizing,* which usually consists of interpreting, defining, naming, outlining, a literary work, movement, genre, or style and providing examples that show how "it" operates in light of pre-existing, theories and concepts, Deleuze rather engages with literary events and in a sense activates them on a new terrain which he configures out of an array of aesthetic, scientific, philosophical, and so on, lines. Deleuze's writing in effect, puts Virginia Woolf's advice into play as he experiments with the literary lines, words, signs of *imperceptible finesse* and the countless names he encounters, to create a new deterritorialized textual terrain, a "*plan of consistence*" or rhizomatic cartography.

The only sort of "literary theory" that can be gleaned concerning Deleuze's work seems to be one that exemplifies a constructivist, creative action/activity in contrast to those that prescribe formulaic procedures based on pre-determined fundamental principles. Deleuze continually expresses his distaste for critical interpretive practices that reduce aesthetic lines, particles and movements to the narrow sphere and practices of discernment tied to "real," concrete phenomena such as, historical or biographical data, and names such approaches "ignoble." (Dialogues, 61). Deleuze's thoughts and writings on "minor" literature, counter an artillery of approaches that close literary meanings and signs in and down in rubrics, grids, and associations linked with pre-determined and determining terms and concepts. Deleuze asserts that it is not enough to speak against such practices, but that one needs to create, write, express and produce thought in a way, that extends multiplicities until the lines that separate into various categories such as, genre, gender, race, and period, become consequently effaced. (Dialogues, 23).

Scherer's article emphasizes this aspect of Deleuze's "theory" or practice that writes *with* literature and other disciplines to move particles, forces and vital signs along, beyond and out of the trappings of "judgmental" interpretations. He does so as he cites and considers Deleuze's adage brought to light in Chapter Two of this text which reads: *« Plutôt être balayeur que juge » "Better to be a sweeper than judge":* This *Deleuzian* line stems from a semantic field that opposes, "finding, meeting, becoming," to "regulating, recognizing and judging": Judges demand conformity, even to rules that they invent. (Dialogues II, 9). Deleuze emphasizes that the traditional French system of philosophy creates a race of *judges* and *legislators* that simultaneously support its foundations as well as those of the "State." This tradition most often paves the way for a kind of philosophizing that nullifies "true thoughts" and instead, imitates, and reinforces oppressive political structures, systems, and institutions. "Ministers of the Interior" and "bureaucrats of pure thought" maintain philosophy's important themes: "an inquiry of understanding," a court of reason, a pure "right of thought": Philosophy becomes the official language of a *Pure State.* "The exercise of thought thus conforms to the goals of the real State, to the dominant meanings and to the expectations of the established order." (Dialogues II, 13)

Two of the vehicles which allow Deleuze to make his way out of the vicious trappings of the French philosophical tradition's "history" and standard practice, are Anglo-American literature and thought. Deleuze emphasizes that not any writing considered literary allows for this release. He points out in countless passages that "cheater" literary artists consciously aim to satisfy the desires of "judges" obsessed with locating the "dirty little secret." These "artists" plant such "secrets" and simultaneously activate and lend credibility to a race of critics and a coinciding interpretive mania. In relation to this Deleuze cites D. H. Lawrence's condemnation of French Literature's craze for the "sale petit secret." "The characters and authors always have a little secret, which feeds the craze to interpret." Deleuze writes that this is why French literature generates manifestos, ideologies, and theories of writing and sets up "narcissistic tribunals." Deleuze prescribes that like minor Anglo-American literature one should never interpret but rather experiment. One needs to construct a "living experiment" moving amongst and joining fragments. Such writings "kill" interpretation. (Dialogues II, 46)

Not only Anglo-American literature provides a way out for Deleuze from systems bent on judging, making points, and interpreting, but its *thought* as well. In Deleuze's later writings especially, the lines separating art and thought blur, contrary to "classical" philosophical approaches. Through this multidimensional way out and the energies he encounters "in flight," Deleuze constructs a deterritorialized terrain that comprises a free flow of lines that run between the traditionally sequestered or separated domains of art and philosophical thought. As Deleuze expresses it in his *Dialogues,* empiricism is the kind of "philosophical" approach that writes, creates, and experiments as the Anglo-American novel. These two domains never interpret, or seek interpretation but rather create strange worlds that continually become and generate further states of flux.

"ON EMIRICISM. Why write, why have written about empiricism, and about Hume in particular? Because empiricism is like the English novel. It is about philosophizing as a novelist, of being a novelist in philosophy. . . . Empiricists are not theoreticians, they are experimenters: they never interpret, they have no principles."(Dialogues II, 54, 55). In the preface to the English translation to *Dialogues* Deleuze refers to himself as an empiricist/pluralist to emphasize that both the empirical philosophical tradition and Anglo-American literature greatly inspire his modus operandi. Above all what his words reveal in this preface is that these two "domains" lead to the articulations of the problems and responses that most concern his work. It can be argued that these problems in accord with Scherer's own insights concern the *mysteries* of the creative process articulated in the questions; "What is it to write? Create? Construct? Desire? And Affirm Life?" "I have always felt that I am an empiricist, that is, a pluralist. . . . The aim is not to rediscover the eternal or the universal, but to find the conditions under which something new is produced (creativeness)." Deleuze continues on to stress why he chooses to think and write as an "empiricist": *Empiricism analyzes the states of things, in such a way that non-pre-existent concepts can be extracted from them. In empiricism "states of things" are neither unities nor totalities, but multiplicities.* Again harkening back to Jean Wahl: "In a multiplicity what counts are not the terms or the elements, but what there is 'between,' the between, a set of relations which are not separable from each other." Deleuze intimately relates Anglo-American literature and empiricism because they both generate these relations, these multiplicities and in doing so respond to the question: 'What is it to write?' with an answer which "is closest to life itself." (Dialogues II, viii, ix).

Gilles Deleuze's words not only remind the reader of his interest in Anglo-American literature and "pragmatic pluralist" constructionist practices, but also reply to remarks that reductively associate the name *Deleuze* with single names which consequently "totalize" or "*total*" his thought. An example of this will lead us back to a quick glance at Deleuze's relationship to, and words concerning, Jean Paul Sartre. To clarify the above issue one need only refer to the example of the contributions that French scholar David Lapoujade, makes to *La Magazine Littéraire's* February (2002) issue dedicated to Deleuze. *La Magazine Littéraire* calls upon Lapoujade to render an accurate résumé of the implications inherent in the principle elements of Deleuze's philosophy. Simultaneously the magazine highlights two of Lapoujade's recent scholarly contributions: One concerns a compilation of Gilles Deleuze's texts that Lapoujade edits, entitled *L'Ile Déserte et Autres Textes,* the other text is entitled, *William James Empirisme et Pragmatisme.* Lapoujade's interpretive expression however might be read as an example of the kind of criticism, theory, *French* way that Gilles Deleuze critiques and turns away from. For Lapoujade sums up Deleuze's philosophy as "classical at heart" indicating Deleuze's homage to Sartre as a sign of his "classicism." What is even more disconcerting is that the aspect of Sartre's thought that Lapoujade aligns Deleuze to is that of "totalization."

> "Deleuze conceived philosophie to be a project of totalization. He makes philosophy with a very classical conception at heart. The homage that he renders Sartre in *Ile déserte,* bears witness to this. One of the things that he admires in Sartre, was this effort of totalization." (Magazine Littéraire, 23)

> « Deleuze concevait la philosophie comme un travail de totalisation. Il se fait de la philosophie une conception très classique dans le fond. L'hommage qu'il rend à Sartre dans *L'île désert* en témoigne. Une des choses qu'il admire chez Sartre, c'est effort de totalisation. » (Magazine Littéraire, 23)

Lapoujade's comments on Deleuze's "conception" of philosophy as one that is "at the heart very classical" must be taken as a gross reduction of Deleuze's project. One must discern that what Deleuze respects and admires in Sartre above all is that despite the fact that he gets stuck in the verb "être," à la philosophie allemande, his entire life he led a revolution against "representing" any one static, reductive ideal as manifest in Sartre's refusal of the Nobel Prize. « Sartre vient de refuser le prix Nobel. Continuation pratique de la même attitude, horreur de l'idée de représenter pratiquement quelque chose, fut-ce des valeurs spirituelles, ou comme il dit, d'être institutionnalisé. » (*Sartre had just refused the Nobel Prize. The continuation of his practice with the same attitude, horrified at the idea of representing almost anything, whether spiritual values, or as he said, to be institutionalized.*) (L'île déserte, 111). For Deleuze it is not Sartre's "system of thought" that is so admirable but Sartre as the model of a voice that refuses being "institutionalized," that never ceases to say things in new, aggressive ways: A voice that vacillates between fiction and philosophy and initiates an interest in crucial writers for Deleuze such as, Kafka, and the American modernists.

> "We know that there is only one value of art and even of truth: that is, "first hand," the authentic newness of what one says, the "petit music" with which one says it. Sartre was that for us (for the twenties generation

at the time of the Liberation). Who then, knew how to say new things if not Sartre? Who taught us new ways to think?. . . . New themes, a certain newness of style, a new polemic and aggressive way to pose problems came from Sartre. In the disorder and the hopes of the Liberation, one discovered, rediscovered it all: Kafka, the American novel, Husserl and Heidegger." (*L'île déserte,* 109, 110)

« Nous savons qu'il n'y a qu'une valeur d'art, et même de vérité: la « première main, » l'authentique nouveauté de ce qu'on dit, la « petite musique » avec laquelle on le dit. Sartre fut cela pour nous (pour la génération de vingt ans à la Libération. Qui alors, su dire quelque chose de nouveau sinon Sartre? Qui nous apprit de nouvelles façons de penser?. . . . Les nouveaux thèmes, un certain nouveau style, une nouvelle façon polémique et agressive de poser les problèmes vinrent de Sartre. Dans le désordre et les espoirs de la Libération, on découvrait, on redécouvrait tout: Kafka, le roman américain, Husserl et Heidegger. » (*L'île déserte,* 109, 110)

Paradoxically, the most decisive argument against Lapoujade's gesture that aligns Deleuze with Sartre's totalizing schema resides in the compilation of Deleuze's texts that Lapoujade himself edits. One asks how Lapoujade could fail to mention that eight years after « *Il a été Mon Maître* » (1964), Deleuze co-authored an article with Michel Foucault « *Les Intellectuels et le Pouvoir* » (1972), that precisely differentiates their philosophical projects from that of Sartre on the almost exclusive grounds of totalization. Deleuze explains that where Sartre theorizes in pursuit of "totalization," they construct fortuitously from fragment to fragment to assemble a non-pre-existing patchwork that never becomes *One, All* or *Total.*

Foucault begins: "*A Mao said to me : Sartre, I understand why he is with us, why he does politics and the way he does them; you to an extent, I understand a bit, you always pose the problem of enclosure. But Deleuze, really a don't understand.*" Foucault continues: "*This question prodigiously surprised me, because it appears very clear for me.*" Deleuze explicitly responds that the difficulty in assessing what "they" do in contrast to Sartre's approach may be resumed as a question of totalization verses multiplication. "One conceived their relationships in the form of a process of totalization, in one way or another. Perhaps for us, the question is raised differently. . . . Theory doesn't totalize, but it multiplies and multiplies itself. Power structures operate totalizations." «*On concevait leurs rapports sous forme d'un processus de totalisation, dans un sens ou dans l'autre. Peut-être pour nous, la question se pose autrement. . . . La théorie, ça ne se totalise pas, ça se multiplie et ça multiplie. C'est le pouvoir qui par nature opère des totalisations.*» (L'île déserte, 288).

> "This is perhaps that we are living the relationships theory-practice in a new way . . . a theory, is exactly as a toolbox. . . . One doesn't come back to a theory, one invents new ones, one has others to make. It is curious that it is an author that passes for a pure intellectual, Proust, who said it so clearly: Treat my book as a pair of glasses directed towards the outside, if they don't work for you, get some others, find your tool which is surely a tool of combat. Theory doesn't totalize it multiplies and multiplies."

> « C'est peut-être que nous sommes en train de vivre d'une nouvelle manière les rapports théorie-pratique . . . une théorie, c'est exactement comme une boite à outils. . . . On ne revient pas sur une théorie, on en fait d'autres, on a d'autres à faire. C'est curieux que ce soit un auteur qui passe pour un pur intellectuel, Proust, qui l'ait dit si clairement : Traitez mon livre comme une paire de lunettes dirigées sur le dehors, eh bien si elles ne vous vont pas, prenez-en d'autres, trouvez vous-même votre appareil qui est forcement un appareil de combat. La théorie, ça ne se totalise pas, ça se multiplie et ça multiplie.» (*L'île déserte,* 288, 290, 291)

Deleuze writes rather as philosopher of the Dehors, the outside, rather than as that of the Total, *unified system.* His theory ever changes and exchanges "tools," in order to continue to construct new, ever open and expanding terrains with lines of flight in a milieu with no beginning, ending or realized, idealized totality. In other words, Deleuze's theory is a contradiction in terms because it activates itself as a mutating, transformative process that alters both itself and that which its practices upon or "reads with," to ever generate, proliferate unique "nomadic distributions" and singularities rather than operating to validate static, staid interpretive practices that coerce works into adjacent frameworks and consequently abort the life forces inherent in acts of creation.

In addition to all that has been said on the issue of interpretation, one finally returns to Lapoujade's *William James Empirisme et Pragmatisme,* full of echoes that resonate out of Deleuze and Jean Wahl's writings to again ponder Lapoujade's gesture which aligns Deleuze's philosophical practice with Sartre's efforts at "totalization" to qualify it as "classical at heart." In Lapoujade's pamphlet like publication of just over 100 pages, themes and terms such as "machine de guerre," "nomad" American workers, *lines* of influence that one extends, to keep the vast network that makes a becoming-world, that makes philosophy a "mosaics philosophy" can not be overlooked. Furthermore Lapoujade fleetingly acknowledges the sources of these terms in two places. On page twenty-eight of the text Lapoujade appeals to Deleuze's words to translate the meaning and particularity of a "transcendental empiricism" that also involves Deleuze's insistence on the role of desire in the creative process that leads to the "construction" of a plan of immanence, or a *radical empiricism.*

> "We borrow the expression from Deleuze who proposes to establish, in *Différence et répétition,* PUF, a "transcendental empiricism." . . ."Transcendental empiricism is (. . .) the only way to not transfer the transcendental on empirical lines." In a completely different perspective, Deleuze invokes a transcendental field without ego, intentionality, uniquely overrun with multiplicities . . . in this sense, the analyses of *Logic of Sense* and *Qu'est-ce que la philosophie?.* . . . The invocation of a radical empiricism, p.49: 'It is when the immanence is no longer immanent in anything other than itself that one can speak of a plan of immanence. Such a plan is perhaps a radical empiricism.'"(Wjames, 28)

> « Nous empruntons l'expression à Deleuze qui propose d'instaurer, dans *Différence et répétition,* PUF, un « empirisme transcendantal. » . . . « L'empirisme transcendantal est (. . .) le seul moyen de ne pas décalquer le transcendantal sur les lignes de l'empirique. » Dans une toute autre perspective, Deleuze invoque un champ transcendantal sans ego, ni intentionnalité, uniquement parcouru de multiplicités . . . à cet égard, les analyses de *Logique de sens* et dans *Qu'est-ce que la philosophie?* . . . l'invocation d'un empirisme radical, p.49: « C'est quand l'immanence n'est plus immanent à autre chose que soi qu'on peut parler d'un plan d'immanence. Un tel plan est peut-être un empirisme radical. » » (WJames, 28)

Further on in his text Lapoujade calls on Jean Wahl to highlight the Jamesean notion of "difference." "The relation of difference is a mélange of continuity and discontinuity. As Jean Wahl states, 'that which is the most contrary to analysis, is not so much the continuum in itself as the apparent mélange of continuum and discontinuum which is the rhythm, or the volume or a person.'" « La relation de différence est un mélange de continuité et de discontinuité. Comme le dit Jean Wahl, *ce qui est le plus contraire à l'analyse, c'est moins le continue en lui-même que ce mélange apparent de continu et de discontinu qu'est le rythme, ou un volume ou une personne.* » (WJames, 64). As Lapoujade, writes with Deleuze, and Wahl, to discern James, he should be well aware that it is precisely *empiricism, pragmatics,* and *pluralism,* the ET verses EST that separates Sartre and Deleuze. Deleuze states clearly that Sartre never gets unstuck from the trappings of the verb "to be" and this is one of the reasons why Sartre, as Deleuze points out, fails in his presentation of "un champ transcendental impersonnel" as he founds it on consciousness that is always a synthesis of unification.

> "We need to determine the impersonal and pre-personal, transcendental field. . . . Despite Sartre's attempt, one cannot maintain the conscience as the center while ignoring the importance of the person and the point of view of individuation. A consciousness is nothing without syntheses of unification, but there is no synthesis of unification of consciousness without the form of the "I" or the point of view of the "Ego." . . . When the world that swarms with anonymous and impersonal, pre-individual nomad singularities opens itself, we will finally be treading upon the transcendental field." (Logique du sens, 124, 125)

> « Nous cherchons à déterminer un champ transcendantal impersonnel et pre-individuel . . . malgré la tentative de Sartre, on ne peut pas garder la conscience comme milieu tout en récusant la forme de la personne et le point de vue de l'individuation. Une conscience n'est rien sans synthèse d'unification, mais il n'y a pas de synthèse d'unification de conscience sans forme du Je ni point de vue de Moi. . . . Quand s'ouvre le monde fourmillant des singularités anonymes et nomades impersonnelles, pre-individuelles, nous foulons enfin le champ du transcendantal. » (Logique du sens, 124, 125).

Furthermore, Sartre explicitly makes it clear in many essays and works that "empiricism" and the "pragmatic" conception of truth, that fail to distinguish between "true" and "false," to yield and support a "systematic" theory of truth in order to "totalize" for the revolutionary cause, operate merely as "subjectivist idealism."

> "The superiority of revolutionary thinking consists in its first proclaiming its active nature; it is conscious of being an act, and if it presents itself as a total comprehension of the universe, it does so because the oppressed worker's scheme is a total point of view toward the entire universe. But as the revolutionary needs to distinguish between the true and the false, this indissoluble unity of thought and action calls for a new and systematic theory of truth. The pragmatic conception of truth will not do, for it is subjectivist idealism, pure and simple." (Literary Essays, 228)

The added examples that help distinguish between Deleuze and Sartre hopefully emphasize that the *traditional French way*, the dialectical course of negativity, more often than not proceeds to designate and interpret at the writer's and his/her work's expense. However despite this, the writings of Gilles Deleuze, as do other "Acts of Resistance," live and pulse on in spite of reductive readings. What is an "Act of Resistance" in Deleuzian terms? Deleuze speaks of "Acts of Resistance" in his televised interview with Claire Parnet entitled "L'Abécédaire de Gilles Deleuze" (Gilles Deleuze's alphabet), as well as in his seminar "Qu'est-ce que l'acte de création?"(May 1987). In the seminar Deleuze insists on a "mysterious" quality inherent in the act of creation or a work of art that qualifies it as an "act of resistance."

> "The work of art absolutely does not contain the least information. On the other hand, on the other hand there is a fundamental affinity between a work of art and an act of resistance. . . . It has something to do with information and communication . . . what is this mysterious relationship between a work of art and an act of resistance?. . . . Malraux says something very simple about art, he says, "This is the only thing that resists death." . . . And what relation between a struggle of men and the work of art? The closest relation is the most mysterious for me. Exactly what Paul Klee wanted to say when he said "You know, the people are missing," that is to say—it isn't clear, it will never be clear—this fundamental affinity between the work of art and a people who don't yet exist. . . . There is not a work of art that doesn't call out to a people who don't yet exist."

> « L'œuvre d'art ne contient strictement pas la moindre information. En revanche, en revanche il y a une affinité fondamentale entre l'œuvre d'art et l'acte de résistance. . . . . Elle a quelque chose à faire avec l'information et la communication . . . quel est ce rapport mystérieux entre une oeuvre d'art et un acte de résistance?. . . . Malraux dit une chose très simple sur l'art, il dit « c'est la seule chose qui résiste à la mort. » . . . Et quel rapport y a-t-il entre la lutte des hommes et l'œuvre d'art? Le rapport le plus étroit et pour moi le plus mystérieux. Exactement ce que Paul Klee voulait dire quand il disait « Vous savez, le peuple manque. » Le peuple manque et en même temps, il ne manque pas. Le peuple manque, cela veut dire que—il n'est pas claire, il ne sera jamais clair—cette affinité fondamentale entre l'œuvre d'art et un peuple qui n'existe pas encore. . . . Il n'y a pas d'œuvre d'art qui ne fasse pas appel à un peuple qui n'existe pas encore. » (Conférence- fondation FEMIS)

In this passage and the words that precede and follow it Deleuze employs the word "mysterious" repetitively to connote an element inherent in the "act of resistance" that exceeds logic or rational explanation. In the televised interview, « *L'Abécédaire* » the letter "R" stands for "Resistance." Deleuze refers to Primo Lévi's writings that recount his experience of Nazi death camps as a form of resistance due to the fact that the writings unleash "forces of life." Parnet objects saying that Lévi's later suicide bears witness to the fact that "art does not suffice." To this Deleuze responds, "He committed suicide personally, he couldn't hold on any longer, so he committed suicide to his personal life. But there are four or twelve pages or one hundred pages by Primo Levi that will remain to constitute an eternal resistance." (Two Fold, 233)

Deleuze constantly distinguishes between the life/lives integral to the writings of thinkers and the life that is associated with their identity or biography. Deleuze writes that despite the fact that some of the greatest thinkers had fragile personal lives and chronic health problems, they were able to transmit "great health" into thought because they carried "life" to an "absolute power." Their creations disseminate resistant forces and allow us to encounter an infinite number of relations, because they give life to a multitude of agencements that incite the circulation of ideas, entities and events: All great thinkers have proper names but these names designate multitudes. (Dialogues, 12, 13)

In *Logique du sens,* Deleuze clearly articulates that the "act of resistance," also refers to a kind of philosophical thought that operates and functions *artistically,* i.e. affirmatively and creatively to promote a *distribution nomade.* This kind of "thought" like art, affirms all chance, fortuity, risk, "le hazard" to advance a nomadic distributive field of play rather than to "*divide and dominate in order to win*" in the manner of traditional philosophical approaches that coerce results to correspond to the expectations of pre-formulated hypotheses and categorical assumptions.

> "Each strike stimulates a distribution of singularities, constellation. But instead of sharing a closed space between fixed results conforming to hypotheses, these are mobile results which spread out into an open space from a unique and non shared launch. . . . Because affirming all chance, makes chance an object of affirmation, only thought can do this. And if one tries to produce a result other than the work of art, nothing is produced. This is therefore, the game reserved for thought and art, here there are only victories for those who know how to play, this is to say to affirm and ramify chance, instead of dividing it in order

to, dominate it, gamble, win. It is also through this game . . . that thought and art are real, and trouble reality, morality and the economy of the world." (Logique du sens, 76).

« Chaque coup opère une distribution de singularités, constellation. Mais au lieu de partager un espace fermé entre des résultats fixes conformément aux hypothèses, ce sont les résultats mobiles qui se repartissent dans l'e-space ouvert du lancer unique et non partagé. . . . Car tout le hasard, faire du hasard un objet d'affirmation, seule la pensée le peut. Et si l'on essaie de produire un autre résultat que l'œuvre d'art, rien ne se produit. C'est donc le jeu réservé à la pensée et à l'art, là où il n'y a plus que les victoires pour ceux qui ont su jouer, c'est à dire affirmer et ramifier le hasard, au lieu de le diviser pour le dominer, pour le parier, pour gagner. Ce jeu . . . il est aussi ce par quoi la pensée et l'art sont réels, et troublent la réalité, la moralité et l'économie du monde. » (Logique du sens, 76).

The *thought* whose "result" corresponds to that of art becomes an "act of resistance" because its affirmative forces are part of a fortuitous field where the "I" and the "personal identity" dissolve to enter an unpredictable series of metamorphoses that Deleuze terms an "intense multiplicity." Deleuze tells us that artistic forces come back to life precisely because they are multiplied in the flow, the field of play and "hazard" (chance), and because once engaged in this zone so many *beings* and *things* think in and through us, to reveal something as powerful as the "pure emotional expression" that reveals "pure motion," "pure *spirit.*"

"Pre-individual and impersonal singularities . . . fascinating World where the identity of the ego is lost, not to the benefit of the identity of the One or the unity of All, but for the profit of an intense multiplicity and a power of metamorphoses. . . . It is not about influences that we experience, but of infiltrations, the fluctuations that we *are,* and those that we confuse ourselves with. . . . Because we are sure to live again (without resurrection) only because so many being and things think in us: because "we don't know if it still isn't the others that continue to think in us-." . . . At the same time bodies lose their unity, and the ego its identity, language loses its function to designate (its manner of integrity) in order to discover a purely expressive value, or, as Klossowski says, "emotional": not in relation to someone who expresses or would be moved, but in relation to a pure expression, pure motion or pure "spirit.""

> « Singularités pre-individuelles et impersonnelles. . . . Monde fascinant où l'identité du moi est perdue, non pas au bénéfice de l'identité de l'Un ou de l'unité du Tout, mais au profit d'une multiplicité intense et d'un pouvoir de métamorphose. . . . Il ne s'agit pas des influences que nous subissons, mais des insufflations, des fluctuations que nous *sommes,* avec lesquelles nous nous confondons. . . . Car nous ne sommes si sûrs de revivre (sans résurrection) que parce que tant d'êtres et de choses pensent en nous: parce que « nous ne savons pas toujours au juste si ce ne sont pas les autres qui continuent à penser en nous. » . . . En même temps que les corps perdent leur unité, et le moi son identité, le langage perd sa fonction de désignation (sa manière à lui d'intégrité) pour découvrir une valeur purement expressive ou, comme dit Klossowski, « emotionelle »: non pas par rapport à quelqu'un qui s'exprime et qui serait ému, mais par rapport à un pur exprimé, pure motion ou pur « esprit. » » (Logique, 346, 347)

As with Primo Levi, Deleuze, or any of the latter's "preferred artists," (Fitzgerald, Woolf, Nietzsche, Kleist, Holderlin), the personal life and its end should have absolutely no bearing on the value of his/her "artistic" creation. For in the ultimate production of *thought* or *art* the "personal" life in a sense renounces itself to both enter and co-create a sphere that entails a dynamic, interaction with and activation of a multitude and multiplication of "pure esprit," entities, becomings, things, and events. "On lines of flight there can no longer be but one thing, experimentation-life. . . . In reality writing does not have its end in itself, precisely because life is not something personal." (Dialogues II, 47). For Deleuze, the objective of writing is to carry the non-personal power of life through and with its movements: To generate a mobile terrain wherein fluxes combine with other fluxes, to allow for "all the minority-becomings of the world." Works that cease to be personal and become a collective force of meetings and conjugations in the process of experimentation both seduce and invite their readers to participate in the creative continuation, to enter and extend the lines, the pulses of flux, to promote their resistance albeit in another form.

Deleuze speaks of his initial works on writers such as Spinoza and Nietzsche as writings of "debt" but the metaphors he uses to describe his encounters with them; *gaining gusts of air, mounting a witch's broom, being taken and released,* suggest an obligation born of receiving, appreciation and "love," *« On n'écrit que par l'amour, toute écriture est une lettre d'amour » (One only writes through love, all writing is a love letter) (Dialogues, 62).* Such experiences move him to *multiply and extend* their works' vital forces beyond the framework of traditional interpretations that prevent the discursive elements of their thought from *functioning.* If an "act of resistance" temporarily misses "people" they will eventually arrive activated and armed with the life forces channeled through them to advance its cause and release it from interpretive gestures that stifle its voices and restrict its movements. "I believe that what concerned me . . . was to describe an exercise of thought, whether in a writer, or for itself, in so far as it opposes itself to the traditional image which philosophy projected of it." (Dialogues, 22, 23)

Concerning all that has been said on "Acts of Resistance," this text considers those writings it has read by, Gilles Deleuze, Jean Wahl, Gertrude Stein, W.E.B. Du Bois, Jean Toomer and William Faulkner as such "Acts" and accordingly has attempted to respond to them. This is to say that all of these texts generated "truths," revelations, a-signifying signs, sounds, realities, that provided what Deleuze describes as "gusts of air," inspiration, infiltration, *des insufflations,* and this text at best hopes it has referenced them, written "with" them in such a way that their meanings breathe on, move on, and ultimately multiply on.

Surely it would be presumptuous and arrogant for a novice in the *Deleuzian way* to claim to write in the manner of Deleuze, to produce a writing within academic confines that succeeds in *writing effectively in such a way that there are no more genres etc., to make multiplicities.* This text claims no such feat however it does consider its movements and encounters, with this variety of writings as propitious nonetheless as they resulted in an experiential and experimental enterprise that indeed involved discovering unsuspected elements fortuitously and non-systematically. This writer is "psychically" convinced that this experience resulted from an exposure to those "resistant" life forces integral to the minor literary, as well as the pluralist, pragmatic texts.

In regards to what has been said on this writing as a sort of apprenticeship or experiment in writing "freely" and freeingly," perhaps that is all such a venture can ever claim to be. Deleuze who describes his own writing as one that over time entered a becoming-rhizomatic-pragmatic enterprise, simultaneously expresses doubt concerning its success at "making multiplicities," and sweeping away categories such as genres and pre-existing conceptual curtailments. "With Felix, all that became possible, even if we failed."(Dialogues II, 16). In the estimation of this writer, Deleuze did succeed in his efforts, to affirm, multiply, create and generate a thought whose results and effects operate as those of art: The innumerable off-shoots and applications of Deleuze's thought today, in cinematic discourse, in musical creations, on poetic websites, confirm this. Again examples of such "applications" mentioned in Chapter One of this text, are the musical/poetic/philosophical creations produced and recently performed internationally by Richard Phinas and Maurice Dantec. One may argue that such "off-shoots" although well-intentioned, this one included, may falsely construe the artist's/philosopher's ideas while putting them into action. However as Rene Scherer points out in another article dedicated to Gilles Deleuze, « *Deleuze éducateur* » all important philosophies must undergo the test of application despite the risk of distortion. For if no one tests, applies, uses a philosophy or its ideas, than it becomes something sterile, dead or worse, the object of veneration or "exegesis" by only the "erudite." According to Scherer, there is no better way to pay homage to Deleuze than to "entrer en scène" with his thought.

> "At the risk of undergoing, heretical interpretations at the hands of users, a philosophy is valued through its use: this is what proves its value and where it also receives its force and fecundity. The deformations that it knows, in passing through stranger's hands, are not as much betrayals as the multiple figures that it contained enfolded in itself, that unfold themselves. These are its metamorphoses, its devenirs. Beyond this, if it doesn't resist this test, it risks turning into something dead, a pure object of exegesis for the erudite, or an object of sterile veneration. Philosophical comprehension is always pragmatic, a putting into action. Also, I am convinced that there isn't a better, more franc method through which to render homage to Gilles Deleuze's philosophy than to use it, each one for his/her own ends."

> « Au risque de subir, de la part de ses utilisateurs, des interprétations hérétiques, une philosophie s'estime à l'usage: c'est là qu'elle prouve sa valeur, qu'elle reçoit aussi sa force et sa fécondité. Les déformations qu'elle connaît, en passant dans des mains étrangères, sont alors moins des trahisons que les multiples figures qu'elle contenait pliées en elle-même, et qui se déploient. Ce sont ses métamorphoses, ses devenirs. Hors cela, si elle ne résiste pas à cette épreuve, elle risque de se transformer en chose morte, pur objet d'exégèse pour érudits, ou de vénération stérile. La compréhension philosophique est toujours pragmatique, une mise en acte. Aussi suis-je convaincu qu'il n'y a pas de meilleure, de plus franche méthode pour rendre hommage à la philosophie de Gilles Deleuze que de la faire servir, chacun à ses propres fins. » (Tombeau, 113)

This reading did begin as an attempt to "use," apply Deleuze's ideas on "minor literature" and its qualities, to American modern texts, however when applying Deleuzian "non-pre-existing concepts" that also resist logic and rational reductions, the efforts to interpret became replaced by a desire to "read and write with," between the art and the literature rather than from one unto the other. One can not describe the transformation of this initial objective as a defeat, retreat, but as something that became a reciprocal operation to result in the construction of an involute, pragmatic space. On/in this space, the lines that should have separated the fields of the literary and the theoretical blurred and dissolved to run into one another, inform one another, and provide vistas into the signs and lives of the other. In other words, this writing did set off to read what it esteemed as American *minor* literary texts through Deleuze's words on adjacent topics such as minoritary-becomings, micro-politics, imperceptible signs, lines of flight, and the like. However, in spite of this aim, as much as Deleuze and his lines helped bring out certain meanings of the literary texts, even more so did the readings, writings on the literature with Deleuze give way to a certain revelation of what the Deleuzian "hints," "signs" of "almost imperceptible finesse" meant "*actively.*" In effect, while Deleuze's textual application led into certain spaces, sights and sounds of the American minor texts, they in turn, and all the other lines this text intercepted, combined to give a "sympathetic" sense of Deleuze's rendering of the creative process, of the way of the "rhizome."

Even if this reading/writing at times makes counter-meanings out of these "beautiful books," and makes their "strange languages," maneuvers, signs say something different than their initial intent, those initial meanings will nonetheless move forward unadulterated. This is because this text never imposes an interpretive dictum, a final revelation, based on pre-determined goals or laws. In light of this, this reading/writing affirms itself as a generative reading that at the very least makes suggestions while allowing the lines it meets, to move on, to fly, escape, disappear, multiply, interconnect, to expand, and proliferate their life forces, energies and entities beyond its scope. Gilles Deleuze, like Rene Scherer, also affirms the beauty of "counter meanings" if generated with creative intentions while he refers to the remarks of Marcel Proust: Because a piece of great literature is written in a sort of foreign language, one often mistranslates it, attaching one's own mental image or meanings to it. *But in great literature all these mistranslations result in beauty.* Deleuze reinforces this to praise mistranslations provided they multiply the book's use and that they do not interpret. (Dialogues II, 5)

As far as the contribution of this text to any debates that vie for the superiority of one critical approach over others, perhaps this text stands simply as an example of an alternative way of reading. Although this way refrains from making "Major" points, and from summing up any ultimate meaning of a work of art, it nonetheless, due to its *demarche,* releases what it deems to be the most vital aspects of the literary and philosophical texts it "reads," that other approaches bent on "making points" usually stifle, overlook, or efface to at least some degree. In addition, this reading may operate and be regarded as a "rhizomatic plan" that discovers and affirms a counter genealogy, not one based on *localizable facts* supported by historical data, nationalized "neighborhoods," geographical boundaries or DNA tests, but rather on "Outside" phenomena that one may glean if willing to allow "le hazard," (*chance*), to play a part in the interpretive process to transmit and reveal lines, grains, particles, gusts of air that "inseminate" psychically rather than physically, that ignite thought processes, insights, and put them "in flight." Such interactions ("distribution nomade"), take place in the *chaosmosphere* or "rhizosphere" in the outer limits of the constellation. "Each strike proliferates a distribution of singularities, constellation. . . . These are all thoughts that communicate in a long thought, that makes all the forms or figure of the nomad distribution correspond through its displacement, everywhere inflating chance." (Logique, 76). One can perhaps understand these ideas more "concretely" if one reads Deleuze's reflections on the illogical connections between thinkers such as Lucrece, Spinoza, Hume, Nietzsche, Bergson. These thinkers "logically," historically, should have nothing to do with one the other yet in the *rhizosphere* they do. Deleuze expresses that something "happens between them" in an ideal, a-historical space. It is not a dialogue of the dead but an "interstellar conversation" that allows their diverse becomings to meet to form a "mobile block." (Dialogues, 22)

In light of what has been said on the rhizomatic way that uncovers a web of "Outside," "distributive nomad" connections verses those made in other interpretive terrains that base their approaches and findings on "historical" data, on pre-formulated ideas that establish what kinds of aesthetic, and ideological products should be "responsibly," "logically" grouped together: This text claims to have shown how a diverse array of texts, of various genres, disciplines, races and nationalities, that conventionally might not ever appear together on one textual space, do indeed have something to do with one the other.

In other words, in a majority of interpretive writings, a novelette written by a 19/20th century expatriate, avant-garde Jewish, homosexual female writer would have nothing to do with a 19th Century Black social scientist's "sociological treatise" sprinkled with song and story. Neither of these would have anything to do with a southern regional writer that concentrates on exposing the psychological drama incurred by both black and whites alike in the post-Civil War South. And, what might any of these have to do with a text hailed the "Harlem Renaissance's" master work that defies any efforts to categorize it in terms of genre. Furthermore, how many scholars, or academic writings, acknowledge, are aware of, or bring out the intricacies related to Gilles Deleuze's wide range of "kinships" with individuals such as Jean Wahl, William James, D. H. Lawrence, Herman Melville, Marcel Proust, Franz Kafka, Virginia Woolf, F. Scott Fitzgerald, or William Faulkner, and the list goes on? The reason that many interesting connections and discoveries are never considered or made amongst various works, artists or philosophers is due to the fact that the majority of interpretive gestures begin with the most obvious, easily accounted for associations that make it possible to make predictable points. Unfortunately many traditional critical approaches more often than not snuff out a work's vital signs, lines and lives as they biologize, historicize, geographize, and separate art and philosophy, to coerce writings and thinkers into predictable, pre-existing scenarios of significance, "relation" and identity. Contrary to this, if one moves in synch with the advice of William James, Jean Wahl and Gilles Deleuze, as Bergson's experimenter ready to enter into and extend the "devenir réel," *A la va comme je te pousse;* "to place oneself at a bound inside the living, moving, active thickness of the real," "*in the making* by a stroke of intuitive sympathy"; a plethora of encounters and exchanges take place, and *truths,* realities, relations, *before the letter* may be received and rendered visible, audible and communicable from within the *between* spaces of the interwoven fabric, the "patchwork," a veritable "plan(e) of immanence."

> "Their names, to be sure, cut them into separate conceptual entities, but no cuts existed in the continuum in which they originally came." (Pluralistic Universe, 285)

> "The rhizome is an antigenealogy. It is a short-term memory, or anti-memory. . . . The rhizome pertains to a map that must be produced, constructed, always detachable, connectable, reversible, modifiable, with multiple entryways, exits, with its lines of flight. Against centered (even polycentric) systems of hierarchical modes of communication and preestablished liaisons. . . . The book as an agencement with the outside, against the book as image of the world. One writes history, but one has always written it from the sedentary's point of view, and in the name of a unitary State apparatus. . . . What is lacking is a Nomadology, the opposite of a history. . . . The nomads invented a war machine in opposition to the State apparatus. History has never understood nomadism; the book has never understood the outside." (Mille Plateaux 32, 34, 35, 36)

Notes

NOTES TO THE INTRODUCTION

1. In *Kafka Pour une Littérature Mineure,* Deleuze makes it clear that minor, or "rhizomatic" texts stimulate a "micro-politics" through their revolutionary molecular movements and forces, such as those proliferated via minor language productions, and collective enunciations, that destabilize the linguistic and legislative, foundations of the "major order." These texts were again chosen for these "revolutionary" or pragmatic (micro-political) aspects inherent in them because as Deleuze states: *It is "la pragmatique" that is essential, because it is true politics and the micro-politics of language.* (Dialogues, 138).

    "The third character is that everything takes on a collective value. . . . The conditions are not given through an *individualized enunciation,* which would be that of such and such "master," and could be separated from the *collective enunciation.* . . . What the writer says by himself already constitutes a common action, and this which he/she says or does is necessarily political. . . . It is literature that is positively obliged in this role and the function of the collective and even revolutionary enunciation: it is literature that produces an active solidarity, despite skepticism; and if the writer is in the margins of or separated from, his/her fragile community, this situation gives him/her even more position through which to express an other potential community, to strike out new means of an other consciousness and an other sensitivity."

    « Le troisième caractère, c'est que tout prend une valeur collective. . . . Les conditions ne sont pas données d'une *énonciation individuée,* qui serait celle de tel ou tel « maître, » et pourrait être séparée de *l'énonciation collective* . . . ce que l'écrivain tout seul dit constitue déjà une action commune, et ce qu'il dit ou fait est nécessairement politique. . . . C'est la littérature qui se trouve chargée positivement de ce rôle et de cette fonction d'énonciation collective, et même révolutionnaire : c'est la littérature qui produit une solidarité active, malgré le scepticisme; et si l'écrivain est en marge ou à l'écart de sa communauté fragile, cette situation le met d'autant plus en mesure d'exprimer une autre communauté potentielle, de forger les moyens d'une autre conscience et d'une autre sensibilité. » (Kafka, 30–33)

2. Brian Massumi who translates many of Deleuze's texts into English points out that Deleuze repetitively refers to his philosophy as a "pragmatics." "Deleuze's own image for a concept is not a brick, but a "tool box." He calls his kind of philosophy "pragmatics" because its goal is the invention of concepts that do not add up to a system of belief . . . but instead pack a potential in the way a crowbar in a willing hand envelops an energy of prying. The question is not: is it true? But: does it work? What new thoughts does it make it possible to think? What new emotions does it make it possible to feel? What new sensation and perceptions does in open in the body?" (Mille Plateaux xv.) The citation from Massumi's introduction to the English addition of *Mille Plateaux,* brings out the most common meaning of "pragmatic" which refers to something's use value or practicality. Although Deleuze does believe a philosophy should and must be used and combined with other elements to generate new ideas, this text will make it clear that Deleuze's insistence on the term "pragmatic(s)" also resonates intimately with ideas associated with the terms "rhizome," "minor-language," and the construction of a "plan of immanence" which ultimately have a "micro"- political effect. In *Dialogues* Deleuze writes: There is no function or organ or corpus of language, but rather *machinic functionings* with collective agencements. 1. It is pragmatics which is essential, because it is the veritable politics, the micro-politics of language. Literature, THE PEOPLE'S AFFAIR, why the most solitary can he say that, Kafka?" On page 22 of a *Thousand Plateaus,* Deleuze even equates these terms writing, "Rhizomatic= Pragmatics= Micropolitics." On page 12 of a *Thousand Plateaus* he expresses: "A rhizome is altogether different, a map and not a tracing. Make a map, not a tracing. . . . What distinguishes the map from the tracing is that it is entirely oriented toward experimentation in contact with the real. . . . It is itself a part of the rhizome. The map is open and connectable in all of its dimensions; it is detachable, reversible, susceptible to constant modification . . . conceived as a work of art, constructed as a political action or as a meditation. . . . The coordinates are determined not by theoretical analyses implying universals but by a pragmatics composing multiplicities or aggregates of intensities." (Thousand, 12, 15)

David Lapoujade who writes, *William Empirisme et Pragmatisme,* which was influenced in part by the writings of Deleuze, also underlines that this "pragmatic" utility necessarily demands an activity of creation that involves constructing a "plan (e)" or a "carte" (map). « *Ce qu'il faut, c'est que les moments fugitives décrits plus haut puissant constituer un véritable plan de construction.* » (What is necessary is for the fugitive moments described above can constitute a true plan(e) of construction.) (Wjames, 11, 21).

3. Deleuze takes this expression that he uses to describe Sartre and other inspiring forces and aesthetic movements such as writing, from a poem written by Bob Dylan. "Sartre was our Outside, he was really the breath of fresh air from the backyard." Deleuze includes the poem in full (modifying the French translation); in the first section of *Dialogues* on pages 13 and 14. Deleuze describes Bob Dylan's poem as proud, marvelous, yet also modest. "It says it all." Deleuze states that he would like to organize a course as Bob Dylan does a song: A lengthy preparation, but without method, rules or recipes. It sweeps rather than judges. According to Deleuze the same principle should apply in philosophy; there are not any "just ideas," there are *just* ideas, but these ideas are the becoming, the "between two," the flight and engagement. (Dialogues 14, 15)
4. David Lapoujade who edits the compilation of Deleuzian texts entitled *Ile Desert et d'autres Textes,* conveys his initiation to a Jamesian pragmatism, or radical empiricism via the writings of Jean Wahl, and Gilles Deleuze, in this book *William James Empirisme et Pragmatisme* and again as far as Deleuze is concerned, in the article entitled, "From Transcendental Empiricism to Worker Nomadism: William James" (2000).

   Lapoujade's remarks on Rorty's particular brand of "neo-pragmatism" as being "ethno-centric" remain true to Deleuze's pragmatic demarche which vies for a move away from all "centers" that distinguish according to myths that apply essentialist ideals of authenticity to identities tied to presuppositions concerning ethnicity, race, and assigned social groupings.

   > "*It isn't about anything other than defining the conversation as an extension of autochthony, a sort of imperialism of Western opinion as the unique source of values. Rorty's recidivated ethnocentrism is in profound contradiction with the pluralism inherent in pragmatism as much as the seeking for consensus with the creative demarche which pragmatism hails as its method. One finds often worrisome characters of "ethnos" in Rorty's texts, that concern a recognition between the representatives of a same community of thought.*"

   > « Il s'agit ni plus ni moins que de définir la conversation comme une extension de l'autochtonie, une sorte d'impérialisme de l'opinion occidental comme source unique des valeurs. L'ethnocentrisme- revendiqué- de Rorty est en contradiction profonde avec le pluralisme inhérent au pragmatisme non moins que la recherche du consensus avec la démarche créatrice

dont le pragmatisme se veut la méthode. On retrouve dans les textes de Rorty les caractères parfois inquiétants de l'ethnos, de cette reconnaissance entre soi des représentants d'une même communauté de pensée. » (Wjames, 123)

5. About French fear and hostility towards American culture and civilization see pages 1–10 of Jean-Philippe Mathy's *Exteme Occident.*
6. David Lapoujade, whose, *William James,* was also influenced by Jean Wahl's writings on Jamesean pluralism, similarly underlines the "democratic" aspect of James's pragmatics. (16, 107). "James's thought is always defined as pluralism, as perspectivism, this is for each consciousness, taken in itself, that the question is raised: how to believe and act/respond? In this sense, the pragmatic method can be defined rightfully as "democratic." It can not dictate any universal rule." « La pensée de James s'est toujours définie comme un pluralisme comme un perspectivisme. C'est pour chaque conscience, prise en elle-même, que se pose la question : comment croire et agir ? En ce sens, la méthode pragmatique peut se définir à bon droit comme « démocratique ». Elle ne peut dicter aucune règle universelle. » (Wjames, 16)
7. In his essay, "Bartleby; or, the Formula," included in *Essays Critical and Clinical* (*Critique et Clinique*), Deleuze makes an interesting reference to "democracy" as an ideal process that creates authentic brotherhood through "originals" that generate an archipelago-perspectivism, an activity that happens in American literature. This link between "democratic" contributions of American literature is again intricately linked to "Pragmatism," and very specifically to that of William James. "Pragmatism is misunderstood when it is seen as a summary philosophical theory fabricated by Americans. On the other hand, we understand the novelty of American thought when we see pragmatism as an attempt to transform the world, to think a new world or new man insofar as they *create themselves*. . . . It is first of all the affirmation of a world in *process,* an *archipelago* . . . not a uniform piece of clothing but a patchwork. . . . But to reach this point, it was necessary for the knowing subject, the sole proprietor, to give way to a community of explorers . . . who replace knowledge with belief, or rather with "confidence." . . . Pragmatism is this double principle of archipelago and hope. ("These themes are to be found throughout Pragmatism, and notably among William James's most beautiful pages: the world as "shot point blank with a pistol.") (Note, 193).
8. Ralph Ellison underlines the fallacy of such presumed, identity associations drawn between race, region or ethnicity and culture in many of his essays. In "The Seer and the Seen," he writes: "Whatever the efficiency of segregation as a socio-political arrangement, it has been far from absolute on the level of culture. Southern whites cannot walk, talk, sing, conceive of laws or justice, think of sex, love, the family or freedom without responding to the presence of Negroes. Similarly, no matter how strictly Negroes are segregated socially

and politically, on the level of the imagination their ability to achieve freedom is limited only by their individual aspiration, insight, energy and will. Wright was able to free himself in Mississippi because he had the imagination and the will to do so. He was as much a product of his reading as of his painful experiences. . . . The same is true of James Baldwin, who is not the product of a Negro store front church but of the library, and the same is true of me." (Shadow, 116)

9. Use of the term, "Major history," has to do with Deleuze's comments that the only history is that of the "majority," a group which has nothing to do with the greatest percentages but with upholding the standards of power structures. The majority imposes its historical perspective on minority groups, and defines them in relationship to the molar system's adjacent hierarchical, binary, organizational grids. Contrarily the "rhizome" generated through minoritarian-becomings and minor languages, always activates a counter-history or histories, and proceeds through an anti-memory process due to the fact that "standard," communal memory is always programmed by the major powers of the "State." "One writes history but one has always written it from the sedentary point of view, and in the name of the unitary apparatus of the State . . . the rhizome is a an anti-genealogy, it is a short term memory, or an anti-memory." (Mille Plateaux, 32).

   "On the contrary to history, "le devenir" (becoming) can not be thought of in terms of past and future. A "devenir- révolutionnaire" (a revolutionary-becoming) remains indifferent to questions about the future and the past of the revolution; it passes between the two. Every becoming is a block of coexistence. The so-called societies without history, place themselves outside of history because they are societies of becoming. . . . There is only history as that of the majority, or of minorities that are defined in relation to the majority." (Mille Plateaux, 358). « Contrairement à l'histoire, le devenir ne se pense pas en termes de passé et d'avenir. Un devenir- révolutionnaire reste indifférent aux questions d'un avenir et d'un passé de la révolution; il passe entre les deux. Tout devenir est un bloc de coexistence. Les sociétés dites sans histoire se mettent hors de l'histoire parce que ce sont des sociétés de devenir. . . . Il n'y a que d'histoire que de majorité, ou de minorités définies pas rapport à la majorité. »

10. The term "hyper-space" is taken from a description that the Italian Poet Andrea Zanzotto uses to describe the cinematographic strategies of Pier Paolo Pasolini. Beverely Allen also uses this term and similarly links it to Pasolini's choice to write his poetry in Friulian which stands as an example of Pasolini's distrust of the dominant language's "power-pedagogies" that infest the larger social-cultural consciousness and unconsciousness. To cite Allen, what Passolini called "style"- the formal or "linguistic," aspects of any work, almost of any gesture- was for him the most efficacious means of opposing this conditioning : "style" verses power pedagogy." Pasolini's

emphasis on stylistic gestures whether linguistic of cinematic is "a constant reminder that the standard language signs are also the signs of oppression, signs never to be completely trusted, signs of power."(SIR, 1). In light of Passolini's aesthetic investment in undermining the oppressive linguistic, semantic or semiological oppressive "pedagogical" vehicles of society helps clarify Deleuze's own reference to Pasolini's "theories" on language as an example of what he means by making a major language's constants shift towards a becoming-minor language, as an attempt to undercut the solidity of the foundations of "Major power structures," to find a way out, and generate a new deterritorialized terrain, a "hyper-space."

"The same goes for minor languages: they are not simply sublanguages, idiolects or dialects, but potential agents of the major language's entering into a becoming–minoritarian of all of its dimensions and elements. . . . That is why Pasolini demonstrated that the essential thing, precisely in free indirect discourse, is to be found neither in language A, nor in language B, but "in language X, which is none other than language A in the actual process of becoming language B." There is a universal figure of minoritarian consciousness as the becoming of everybody, and that becoming is creation. . . . It is certainly not by using a minor language as a dialect, by regionalizing or ghettoizing that one becomes revolutionary; rather, by using a number of minority elements, by connecting, conjugating them, one invents a specific, unforeseen, autonomous becoming." (Thousand, 106)

11. Virginia Woolf, one of Deleuze's most cited authors, equally exposes the contestable assumption that the reader or critic's own personal prejudices and world perspectives that construct his/her visions of reality are more "truthful" than the world and language generated in a work of art.

"But here many difficulties arise. For we have our own vision of the world; we have made it from our own experience and prejudices, and it is therefore bound up with our vanities and loves. It is impossible not to feel injured and insulted if tricks are played and our private harmony is upset. Thus when *Jude the Obscure* appears or a new volume of Proust, the newspapers are flooded with protests. Major Gibbs of Cheltenham would put a bullet through his head tomorrow if life were as Hardy paints it; Miss Wiggs of Hampstead must protest that though Proust's art is wonderful, the real world, she thanks God, has nothing in common with the distortions of a perverted Frenchman. Both the gentleman and the lady are trying to control the novelist's perspective so that it shall resemble and reinforce their own. But the great writer—the Hardy or the Proust—goes on his way regardless of the rights of private property. . . . In masterpieces—books, that is, where the vision is clear and order has been achieved—he inflicts his own perspective upon us so severely that as often as not we suffer agonies—our vanity is injured because the old supports are being wrenched from us." (Vwoolf, 53, 54)

12. A passage from *Dialogues* clarifies what the term "involute" means in a Deleuzian context as it relates to the "Devenir," the milieu, the rhizomatic, pragmatic, process. Deleuze relates "involution" intimately to "le devenir" which involves neither the past, future, nor even the present, as it is divorced from a notion of history. "Becoming is rather a matter of "involuting"; it is neither regression not progression." Becoming through "involution" involves becoming more deserted, simple, and sober and yet that much more "populated." To involute is to be "between," as in Virginia Woolf's promenade amongst the taxis; the stroll as an act of politics, experimentation, and life. "I spread myself out like fog BETWEEN the people that I know the best." (Dialogues, 37)

NOTES TO CHAPTER ONE

1. In *Dialogues,* Deleuze speaks of the 'regime of signs' or agencements, as that which a proper name disperses. This name has nothing to do with a subject or a unified identity but rather generates a mutating and transformative "collective." Deleuze draws an analogy between the names of famous Doctors who study or discover a particular disease, such as Parkinson whose names come to represent a plethora of aspects that relate to the symptoms of the sickness, with the power of the thought associated with one particular thinker. Deleuze asks in addition to these names such as "sadism," "masochism" why there isn't also a "nietzscheisme," "proustisme," "kafkaïsme," and "spinozisme"? (Dialogues, 143)
2. This letter to Michel Cressole also speaks to Deleuze's own interpretation of Foucault's comments, "One day the age will be known as "Deleuzian," and to the element of complicity inherent in their long time friendship while it also reveals Deleuze's wit. "Of course, benevolence is not your strong point. When I am no longer capable of loving and admiring people and things (not very many), I'll feel dead, mortified. But as for you, it seems that you were born sour; all art is in allusions. 'I won't be taken in. . . . I'm writing a book on you, but I'll show you.' Of all possible interpretations you'll generally choose the most wicked or the vilest. First example: I love and admire Foucault. I've written an article on him. And he, one on me, from which you quote the following sentence: 'Perhaps the century will be called Deleuzian one day.' Your comment: they send each other flowers. It seems you can never get the idea that my admiration for Foucault is real and that Foucault's statement is just a crack intended to make those people laugh who love us and to make the others rage."
3. Stivale's *Two Fold,* documents Foucault's shared esteem for the passion and courage inherent in Sartre's intellectual demarche while it also notes that Foucault and Deleuze together discuss their projects as distinct from that of

Sartre emphasizing the major role that Nietzsche plays in initiating their divergence. "During an interview they gave together in July 1966, in which Deleuze and Foucault described their project to reedit Nietzsche's complete works, they contrasted the impact of Nietzsche on contemporary Western thought to the Sartrean legacy. . . . 'Nietzsche opened a wound in philosophical language. Despite efforts by specialists, the wound has never been healed.'" (Two Fold, 285)

4. Charles Guenther, who edits and translates Jean Wahl's poems in the collection, *Voices in the Dark: Fifteen Poems of the Prison Camp,* documents the professor, student relationship that existed between Wahl and Sartre, that continues on in Wahl's commentaries on Sartre's existential writings. "On June 19, 1974, the French poet and existentialist philosopher Jean André Wahl died in Paris at the age of 86. Somewhat less known than Jean-Paul Sartre (whom he failed in a course at the Sorbonne), Wahl left a significant legacy of ideas in philosophy." (Voices, 5)
5. This same book, *La Nausée* by Rhiannon Goldthorpe, also sheds light on the extent to which Sartre did contradict himself in his philosophical affinities especially when one compares the young Sartre's appreciation of Nietzsche with the later Sartre of *Literary Essays.* "Sartre also mentions that while at the Ecole he preached a 'Nietzschean morality of joy, even though, in other respects, all joy and all harshness proved to be impossible in the contingent, nauseating world I had discovered.' (113). "His contemporaries recall that he first formulated his own ideas on contingency while presenting a paper on Nietzsche in 1927 or 1928. Furthermore, during this period he was not only reading Charles Andler's monumental account of Nietzsche's life and philosophy, but also writing a novel inspired by the early relationship between Wagner, Cosima Wagner and the young Nietzsche, with whom Sartre identified himself." (21)

   This account contrasts radically with Sartre's 1955 essay "Departure and Return" in which he critiques Parain's taste for Nietzsche's "crackbrained nonsense." "Parain has not hesitated to reproduce a weak analysis of the Cartesian cogito that he found in *The Will to Power.* We know that Nietzsche was not a philosopher. But why does Parain, who is a professional philosopher, quote this crackbrained nonsense?" (Literary Essays, 171)
6. Ernst Behler emphasizes the decline in Nietzsche's popularity during and after the Second World War due to an association between Nazi propaganda and Nietzsche's philosophy. "During the first half of the century he was one of the most influential and widely read authors, and his name enjoyed a meteoric rise in popularity. During World War II, however, and especially after its end, interest in Nietzsche lessened. Thomas Mann, in a 1947 address to the Pen Club of Zurich entitled "Nietzsche's Philosophy in the Light of Our Experience," described the attitude of that time toward Nietzsche as one of "lamentation" and captured this mood in *Doctor Faustus.*" (Confrontations, 2)

7. This reference to Sartre as "Master" comes from Deleuze's article, « Il A ETE MON MAITRE, » published originally in *Arts,* in November of 1964. While Deleuze does sincerely esteem Sartre, one should not ignore the irony of Deleuze's gesture to so name Sartre whose philosophical demarche never ceases to refer to the Hegelian master/ slave dialectic and to a totalizing system. (Literary Essays, 225). Furthermore, all throughout Deleuze's writings he maintains his abhorrence for Hegelianism. « Ce que je détestais avant tout, c'était le hégélianisme et la dialectique. » (*What I hated above all was Hegelianism and the dialectic.*) (Pourparlers, 14). One sees the predominance of Hegelian themes in *Being and Nothingness* (1943), as well as in Sartre's later works such as, *Materialism and Revolution* (1955). "It is precisely in becoming revolutionaries, that is, in organizing with other members of their class to reject the tyranny of their masters, that slaves best manifest their freedom." (Literary Essays, 245)
8. Wahl also participated in contributions to the journal *Acéphale*. "*Acéphale* was both a secret society and a publication. The magazine was published after *Documents* . . . Bataille's colleagues Pierre Klossowski, Roger Caillois and Jean Wahl contributed. The artist André Masson did much of the artwork. The College of Sociology (*Collège de Sociologie*) was the theoretical counterpoint to *Acéphale,* and included many leading intellectuals of the day—Walter Benjamin, Jean Paulhan, Theodor Adorno, Claude Lévi-Strauss." (Alphabetical List of 20th Century Avant-Garde Periodicals in the Sheridan Libraries Collections John Hopkins's University). http://www.library.jhu.edu/rsd/other/surreal/alpha.html .

   Furthermore, Jacques le Rider affirms that a "young Deleuze" began attending meetings at Marcel Moré's who attempted to keep the "College" going during the war. Frequent participants included, Sartre, Georges Bataille, Jean Wahl, Pierre Klossowski, among other notorious French intellectuals.

   "Marcel Moré, had attempted to save the Collège by organizing meetings at his apartment. . . . Jean Wahl was a faithful participant at this club until his exile to the United States . . . Jean Wahl had even stopped in one more time at Marcel Moré's, one Saturday, while his friends were urging him to leave the occupied zone. . . . Maurice de Gandillac also remembers the Saturday of March 1944, when one discussed Bataille's ideas on Nietzsche, while France went through one of the darkest periods of its history. Sartre, Simone de Beauvoir, Klossowski, Adamov, Danielou, Gabriel Marcel, Léiris, Berdiaev, Queneau, Hyppolite were present. . . . This is also in 1944 that the young Deleuze was introduced into Marcel Moré's circle and took part in its debates." « Marcel Moré, avait tenté de sauver le Collège en organisant des réunions dans son appartement. . . . Jean Wahl fut un fidèle de ce cénacle jusqu'a' son départ en exil aux Etats-Unis . . . Jean Wahl avait tenu à passer encore une fois chez Marcel Moré, un Samedi, alors que ses amis le pressaient de quitter la zone occupée. . . . Maurice de Gandillac garde aussi le souvenir de ce Samedi de mars 1944, où l'on discutait chez Marcel Moré des thèses de Bataille sur Nietzsche, alors que la France traversait une des périodes les plus sombres de son histoire. Sartre, Simone de Beauvoir, Klossowski, Adamov, Danielou, Gabriel Marcel, Léiris, Berdiaev, Queneau, Hyppolite étaient présents. . . . C'est aussi en 1944 que le jeune Deleuze fut introduit dans le cercle de Marcel Moré et associe à ces débats. » (Nietzsche en France, 182, 183)

9. Deleuze intimately links Nietzsche's "Will to power" and 'desire' to his description of the creative, political, activating processes of a "plan de consistence," the construction of this "plan(e)," ultimately stimulates assemblages of "social becomings." Deleuze tells us that in constructing the plan latent desire surges up but one needs to know how to find one's proper directions to map out such a plan. Deleuze insists that this plan and its processes are fueled by "life," by this desire that self-generates in process. This process has nothing to do with those who believe that creations arise from "lack," from resentment. "Desire: who except priests, would want to call it 'lack'? Nietzsche called it 'Will to Power.'" (Dialogues 109, 110/Dialogues II, 90, 91)
10. These kind of racial, essentialist claims that one also finds in Sartre's *Orphée* are the same Sartrean ideas that Franz Fanon, critically responds to in his book, *Black Skin, White Masks:* "Nevertheless with all my strength I refuse that amputation. I feel in myself a soul as immense as the world, truly a soul as deep as the deepest of rivers, my chest has the power to expand without limit. I am a master and I am advised to adopt the humility of the cripple." (Black Skin, 140).
11. Deleuze comments positively on Jean Wahl's remarks at the Nietzsche Colloquium at Royaumont (1964), at which Deleuze gave the closing presentation and concluding remarks. Deleuze emphasizes that Wahl's contribution along with that of Pierre Klossowski, speaks to the inherent plurality, mobility and plasticity of Nietzsche's thought: « Je suis Chambige, je suis Badinguet, je suis Prado, tous les noms de l'histoire au fond c'est moi. » « Mais déjà M. Wahl avait fait le tableau de ce gaspillage génial avant la maladie, de cette mobilité, de cette diversité, de cette puissance de métamorphose qui forment le pluralisme de Nietzsche. » ("I am Chambige, I am Badinguet, I am Prado, all the names of history at the depth are me." But already Monsieur Wahl made the picture of this fabulous excess in the face of illness, of this mobility, this diversity, this power of metamorphosis that makes up Nietzsche's pluralism.") (Cahier, 276)

NOTES TO CHAPTER TWO

1. A passage in James's *Pragmatism,* precisely speaks to the opposition of the distributive, diffused world of *eaches* which only the "tough-minded" can

accept, as this world involves an acceptance of unknowns, insecurity and the instability of particles, "eaches" that don't easily conform to or fit into the "totalizing" structures of the, *All*-into *One*-world of the "tender-minded" that "secures the eaches without exception": In other words that effaces, suppresses or erases the "eaches." "You see how differently people take things. The world we live in exists diffused and distributed, in the form of an indefinitely numerous lot of *eaches,* coherent in all sorts of ways and degrees; and the tough-minded are perfectly willing to keep them at that valuation. They can *stand* that kind of world, their temper being well adapted to its insecurity. Not so the tender-minded party. They must back the world we find ourselves born into by "another and a better" world in which the eaches form an All and the All a One that logically presupposes, co-implicates, and secures each *each* without exception." (Pragmatism, 102)

2. This emphasis on the "And" resonates with James's passage from a Pluralistic Universe. One can make the case that what Deleuze admires so much in Wahl is very "Jamesean." "I now say that the notion of 'one' breeds foreignness and that of the 'many' intimacy, for reasons which I have urged at only too great length, and with which, whether they convince you or not, I may suppose that you are now well acquainted. But what at bottom is meant by calling the universe many or by calling it one? Pragmatically interpreted, pluralism or the doctrine that it is many means only that the sundry parts of reality *may be externally related.* . . . Things are 'with' one another in many ways, but nothing includes everything. The word 'and' trails along after every sentence. Something always escapes." (Pluralistic, 321).
3. In Du Bois's *Writings* he emphasizes the distinction between American prejudicial attitudes towards black Americans and the more humane treatment African Americans received in France during the "Great War." He also relays accounts of a propaganda campaign that was to encourage the spread of fear and discriminatory attitudes towards African American soldiers amongst the French but was overwhelmingly unsuccessful. Furthermore, this essay attributes the main surge of "dogged determination" amongst blacks to work for equality on American soil as having been incited by their experience of France's truer, color- blind democratic values. Taking this into account one must consider the "Harlem Renaissance" movement as an integral result of these historical experiences, as an extension of the energies of this new determination for change that in part takes form in aesthetic expressions, to ignite a new kind of political battle. "Little was published or openly said, but when the circular on American Negro prejudice was brought to the attention of the French ministry, it was quietly collected and burned. And in a thousand delicate ways the French expressed their silent disapprobation. . . . For the Negroes this double experience of deliberate and devilish persecution from their own countrymen, coupled with a taste of real democracy and world-old culture, was revolutionizing. They began

to hate prejudice and discrimination as they had never hated it before. They began to realize its eternal meaning and complications. Far from filling them with a desire to escape their race and country, they were filled with a bitter, dogged determination never to give up the fight for Negro equality in America. . . . A new, radical Negro spirit has been born in France, which leaves us older radicals far behind." (Writings, 883, 894–5).

4. In *Black Atlantic,* Paul Gilroy emphasizes the auditory aspects of *Souls,* its, 'polyphonic, montage technique,' which becomes a precursor to works by artists such as James Weldon Johnson and of course one can see the influence of *Souls* in the arrangements of Jean Toomer's *Cane.* "Souls is a key text. It underpins all that follows it, and its importance is marked by the way Du Bois places black music as the central sign of black cultural value, integrity and autonomy." (Black Atlantic, 89, 90)
5. In *Terrible Honesty: Mongrel Manhattan in the 1920s,* Ann Douglas documents this teacher, student relationship and also documents William James impression of *Souls,* marked by his passing it on to his brother Henry to read. She also underlines William James's investment in ideas that speak to the "pragmatic, pluralistic" potential forces of "popular culture" and the extent to which Jamesian pragmatics encouraged avant-garde cultural productions, and notably those of the Harlem Renaissance movement. "James read Du Bois's *The Souls of Black Folk* in 1903 and passed it on to his brother the novelist Henry James as "a decidedly moving book." Henry believed it "the only 'Southern' book of any distinction published in many a year," (Again concerning William James), "It might be said that he laid the philosophical basis for the American preference for popular culture over elite and self-consciously difficult art, for the choice of culture over politics that gave the Harlem Renaissance its point of origin. His notion of American culture as a plural and heterogeneous affair of simultaneous affects, collaboratively improvised what he called "the will to personate," was a viewpoint congenial to black aims and achievements; what the syncopated black ragtime music of the 1890s and 1900s was to Euro-American classical music, the quicksilver and irregular Jamesian discourse was to traditional Western thought." (Terrible Honesty, 115, 116)
6. In *Pluralistic Universe* James referring to Bergsonian ideas, urges creative, generative action, that runs contrary to an "intellectualist, post-mortem type of criticism." "Get at the expanding center of a human character, the *élan vital* of a man, as Bergson calls it, by living sympathy, and at a stroke you see how it makes those who see it from without interpret it in diverse ways. . . . What really *exists* is not things made but things in the making. Once made, they are dead, and an infinite number of alternative conceptual decompositions can be used in defining them. But put yourself *in the making* by a stroke of intuitive sympathy with the thing and, the whole range of possible decompositions come at once into your possession. . . . Reality *falls*

in passing into conceptual analysis; it *mounts* in living its own divided life- it buds and bourgeons, changes and creates. Once adopt the movement of this life in any given instance and you know what Bergson calls the *devenir réel* by which the thing evolves and grows. Philosophy should seek this kind of living understanding of the movement of reality, not follow science in vainly patching together dead results." (Pluralistic Universe, 262, 264).

7. As does James, Du Bois speaks of the difficulty of expressing certain realities that his *Souls* evokes in conceptual terms. He also refers to the need to break with traditional, authoritarian, narrative style, and "cast off the restraint of (his) training and surroundings." Simultaneously, Du Bois intimates that this stylistic break in part is intertwined with relaying dimensions of an "other" world, "a world as we see it who dwell therein," that world and its adjacent experiences and consciousness is that of "black America": A world that white America remains ignorant of, alienates and shuns.

   "There are bits of history and biography, some description of scenes and persons, something of controversy and criticism, some statistics and a bit of story-telling. All this leads to rather abrupt transitions of style, tone and viewpoint and, too, without a doubt, to a distinct sense of incompleteness and sketchiness. . . . Through all the book runs a personal and intimate tone of self-revelation. In each essay I sought to speak from within- to depict a world as we see it who dwell therein. In thus giving up the usual impersonal and judicial attitude of the traditional author I have lost in authority but gained in vividness . . . some revelation of how the world looks to me cannot easily escape him (the reader)." (Book Reviews By W.E.B. Du Bois, 1903)

8. This "mind blowing inner catastrophe" of course relates particularly to the early 20th century reader as the book was first published in 1903, preceding the Harlem Renaissance movement by almost two decades.

9. This issue of the power and pre-linguistic or supra-logical force of the sorrow songs, resonates with Deleuze's own words on the subject of African American slave or sorrow songs. Deleuze points to these songs as "transformational" in that they explode (faire éclater) standardized, sign regimes on a plan of consistence that involves an absolute deterritorialization of the major language. Deleuze also notes Le Roi Jones in this section and his *Blues People* which speaks to Deleuze's interest in and reflection upon the linguistic, musical, and literary counter-cultural forces of Black American Cultural products. "The songs of Black Americans, especially including the lyrics, would have an even more exemplary value, because one hears first how the slaves "translate" the English signifier, and make a pre-signifying or even counter-signifying use of the language." « Les chansons des Noirs américains, y compris et surtout les paroles, auraient une valeur encore plus exemplaire, parce qu'on y entend d'abord comment les esclaves "traduisent" le signifiant anglais, et font un usage présignifiant ou même contre-signifiant de la langue. » (Mille Plateaux, 171)

10. In addition, it seems that Gibson wants to emphasize that Arthur Symons is of another generation, economic class, social strata than "the creators" of "Nobody Knows the Trouble I've Seen" due to his own stereotyping of a "typical" black American background. However, contrary to this statement, Harry T. Burleigh and Symons were contemporaries as Burleigh was born in 1866 and Symons in 1865. Furthermore, in opposition to this supposed "chasm," social and economic, Burleigh was raised by a college educated mother who spoke French and Greek fluently. He eventually earned a scholarship to the National Conservatory of Music were he worked intimately under the tutelage of Czech composer Antonin Dvorak. "Dvorak came to the United States in 1892 as the new director of the conservatory. He learned of the spiritual through his contacts with Burleigh and later commented that: " . . . inspiration for truly national music might be derived from the Negro melodies or Indian chants. I was led to take this view partly by the fact that the so-called plantation songs are indeed the most striking and appealing melodies that have yet been found on this side of the water, but largely by the observation that this seems to be recognized, though often unconsciously, by most Americans. . . . The most potent as well as most beautiful among them, according to my estimation, are certain of the so-called plantation melodies and slave songs, all of which are distinguished by unusual and subtle harmonies, the like of which I have found in no other songs but those of old Scotland and Ireland." Burleigh became such a prolific composer and performer that he was invited to give concerts for such dignitaries as the King and Queen of England and President Theodore Roosevelt. "Burleigh was a charter member of the American Society of Composers, Authors, and Publishers (ASCAP) when it formed in 1914 and became a member of its board of directors in 1941. He received a number of honors, including the Spingarn Medal in 1917, and honorary degrees from Atlanta University and Howard University for his contributions as a vocalist and composer." (http://www.afrovoices.com/burleigh.html)
11. Deleuze often refers to Black American English, music, dance, linguistic play, to exemplify what he means by the activity of "working on a language" to make elements shift from "major" to "minor" to the point that he states, "American is indeed the Black's language." Furthermore it is striking that this passage is included in the small section of *Dialogues* devoted to "Empiricism" and directly follows Deleuze's homage to Jean Wahl and the "Art of the ET." This inadvertently reinforces the pluralist-pragmatic aspects that function at the heart of Du Bois's text. In Deleuze's words minorities "work upon" the American language and put it into their own service. Deleuze states that if slaves learned standard English, it was in order to "flee" by putting the master's language to flight. In *Mille Plateaux* (129–134) when Deleuze again affirms that American is the language of the "blacks," he also confers that minor authors like minorities "work on the

major language," and make it shift until it becomes "minor." "That is the strength of the authors termed "minor," who are in fact the greatest, the only greats: having to conquer one's own language . . . to place it in a state of constant variation." (Thousand Plateaus, 104, 105).

12. Please see Note 1, Chapter Two again in relation to this section.
13. Du Bois speaks again of a distinct clairvoyance particular to Black Americans regarding America in his 1926 essay, "Criteria of Negro Art." "Once in a while through all of us there flashes some clairvoyance, some clear idea, of what America really is. We who are dark can see America in a way that white Americans cannot. And seeing our country thus, are we satisfied with its present goals and ideals?" (Writings, 993).
14. Again in the essay, "Criteria of Negro Art," Du Bois writes of the aesthetic as a potential venue through which to "find a way out" as a weapon through which to efface the "color line." "With the growing recognition of Negro artists in spite of the severe handicaps, one comforting thing is occurring to both white and black. They are whispering, "*Here is a way out.* Here is the real solution of the color problem. The recognition accorded Cullen, Hughes, Fauset, White and others shows there is no real color line. Keep quiet! Don't complain! Work! All will be well!"(Writings, 997) (My emphasis)
15. Du Bois even writes twenty-three years after his *Souls,* about the trials of being published as a Black writer if what one writes accurately represents "black life," beyond the stereotypes of the minstrel like figure: the "happy, comical, simple-minded 'darky.'" Despite this Du Bois still urges for Black writers to render authentic portrayals of black life and culture in all of its diversity as again he emphasizes that this may be the one way through which to move individuals to a color blind state of psychology, to push individuals finally to the question : "What is a Negro anyway? "Suppose the only Negro who survived some centuries hence was the Negro painted by white Americans in the novels and essays they have written. What would people in a hundred years say of black Americans?." . . . Suppose you were to write a story and put in it the kind of people you know and like and imagine. You might get it published and you might not. And the "might not" is still far bigger than the "might." The white publishers catering to white folk would say, "It is not interesting"—to white folk, naturally not. They want Uncle Toms, Topsies, good "darkies" and clowns. . . . We must come to the place where the work of art when it appears is reviewed and acclaimed by our own free and unfettered judgment. . . . And then do you know what will be said? It is already saying. Just as soon as true Art emerges; just as soon as the black artist appears, someone touches the race on the shoulder and says, "He did that because he was an American, not because he was a Negro; he was born here; he was trained here; he is not a Negro-what is a Negro anyhow?" (Du Bois Writings, 999, 1001, 1002).

16. This reference to the "same voice" that sings both one of the spirituals and a German folk song, relates to a very interesting passage in the chapter the *Sorrow Songs,* where Du Bois interrelates two seemingly dissimilar songs, again through a disembodied non-raced, non-nationalized voice. On page 212 one first reads three bars of music, matched with lyrics that read, "Poor Rosy, Poor gal; Poor Rosy, poor gal, Rosy break my poor heart. Heav'n shall be my home." Then the text beneath reads, "A black woman said of the song, "It can't be sung without a full heart and a troubled sperrit." The same voice sings here that sings in the German folk-song: "Jetz Geh i' an's brunele, trink' aber net." (Souls, 212)

NOTES TO CHAPTER THREE

1. In *Mille Plateaux* Deleuze distinguishes between minoritarian (minoritaire) and majoritarian (majoritaire). The latter is a constant and homogeneous system, the minoritarian rather is a creative potential, a "becoming-potential." Through "minoriarian" movements and creations, the "Nobody," has a chance to enter an "everybody"-becoming, or a becoming- "everybody": a becoming that deviates from the standard model of Being within the "majority" order. Within this passage the distinction between "minority" and "majority" is also made. It becomes clear that in a Deleuzian field of meaning, what differentiates these two terms has less to do with numbers than with the latter, "representing a constant," reinforcing the stability of the "Major Molar order."

   "The opposition between minority and majority is not simply quantitative. Majority implies a constant, of expression or content, serving as a standard measure by which to evaluate it. Let us suppose that the constant or standard is the average adult-white-heterosexual-European-male-speaking a standard language (Joyce's or Ezra Pound's Ulysses). It is obvious that "man" holds the majority, even if he is less numerous than mosquitoes, children, women, blacks, peasants, homosexuals, etc. That is because he appears twice, once in the constant and again in the variable from which the constant is extracted. Majority assumes a state of power and domination, not the other way around. It assumes a standard measure, not the other way around. . . . A determination different from that of the constant will therefore be considered minoritarian, by nature and regardless of number, in other words a subsystem or an out-system (hors système). . . . For the majority, insofar as it is analytically included in the abstract standard, is never anybody, it is always Nobody-Ulysses- whereas the minority is the becoming of every-body. That is why we must distinguish between: the majoritarian as a constant and homogeneous system; minorities as subsystems; and the minoritarian as a created and creative, potential becoming (devenir). The problem is never to acquire the majority, even to install a new constant. There is no becoming-majoritarian; majority is never becoming. All becoming is minoritarian." (Thousand, 105, 106/Mille Plateaux, 133, 134)

2. Ann Douglas's book, *Terrible Honesty,* marks the influence William James had on Gertrude Stein when she was a student of his at Harvard in the min 1890s, "The two at once admired and appreciated each other; they continued to do so until there respective deaths in 1910 and 1946." (Terrible, 119). Stein refers to James as one of her "forerunners," "because he was the only 19th century writer who was so "American" that he "felt the method of the twentieth century." "To be American, Stein's logic went, is to be modern, irrespective of dates; the more American, the more modern." (Terrible, 118)
3. Deleuze relays that racism is always activated in a field of opposition between the "Standard" symbol or representative of the Majority, its order and language, which is the White-European- heterosexual-male speaking the standardized major language. (Note 1/Chapter Two). Due to this absence of both standardized English and "standardizing" white male, one may more easily detect the autonomous, minoritarian, particles of the becoming-others-everybody, "minoritarian others" that conjugate with and connect variations of minority elements: that diverge from homo-sexual, homo-linguistic, homo-racial. "Racism operates by the determination of degrees of deviance in relation to the White-Man face, which endeavors to integrate nonconforming traits into increasingly eccentric and backward waves, sometimes tolerating them at given places under given conditions, in a given ghetto, sometimes erasing them from the wall, which never abides alterity (it's a Jew, it's an Arab, it's a Negro, it's a lunatic . . .). From the point of view of racism, there is no exterior, there are no people on the outside. There are only people who should be like us and whose crime it is not to be. The dividing line is not between inside and outside but rather is internal to simultaneous signifying chains and successive subjective choices. Racism never detects the particles of the other; it propagates waves of sameness until those who resist identification have been wiped out (or those who only allow themselves to be identified at a given degree of divergence). Its cruelty is equaled only by its incompetence and naïveté." (Thousand, 178)
4. Deleuze writes that "becomings" "belong to geography; they are orientations, directions, entries and exits." The "woman-becoming" has nothing to do with the woman's past and their future but the woman must enter this becoming to get out of her past, future and history. However it is not necessarily the woman that activates the devenir-femme or militants who initiate a revolutionary-becoming. In the same vane philosophy-becoming has nothing to do with the history of philosophy. It happens through those whom the *history of philosophy fails to classify.* (Dialogues, 2)
5. One can say that Melanctha like Bartleby is an "original." A character that defies categories, and logic, exceeds rational or psychological explanations

or theories to throw a glaring light on the imperfection of the world's often inconsistent laws and its orders that encourage and reward lives of "mediocrity." "Figures of life and knowledge, they know something inexpressible, live something unfathomable. They have nothing general about them, and are not particular- they escape knowledge, defy psychology. Even the words they utter surpass the general laws of language (presuppositions) as well as the simple particularities of speech, since they are like the vestiges or projections of a unique, original language (langue), and bring all of language to the limit of silence and music. . . . The original, says Melville, is not subject to the influence of his milieu; on the contrary, he throws a livid white light on his surroundings." (Bartleby or the Formula, 83)

6. *Melanctha's* resistance to psychological explanations, her transgression or subversion of moral codes and laws and Stein's refusal to unveil the reasons or secrets behind Melanctha's complexity and contradictory acts and nature should be considered in light of Gertrude Stein's rejection of psychoanalysis, and her pragmatic experimental demarche that puts Melanctha's wisdom building, wanderings into action. "The English novel, and even more the French novel, feels the need to rationalize, even if only in the final pages, and psychology is no doubt the last form of rationalism: the Western reader awaits the final word. In this regard, psychoanalysis has revived the claims of reason. . . . The founding act of the American novel, like that of the Russian novel, was to take the novel far from the order of reasons, and to give birth to characters who exist in nothingness, survive only in the void, defy logic and psychology and keep their mystery until the end." (Bartleby or the Formula, 81)
7. Interestingly enough, Ann Douglas underlines that both James and Stein believed in an ability to "modify" the unconscious through conscious movements, and one can see how this Anti-Freudian stance for which Stein was famous, took shape in her aesthetic practice. "James and Stein seem to be running on shared insights, improvising a collaboration. There is in both an instinctive reliance on the conscious life to express and modify the unconscious one. . . . Both are quick to privilege a certain kind of feminine over the note-taking masculine mind, and they are wonderfully willing to drop the rules at the call of experience."(Terrible, 119) In *Dialogues,* Deleuze also articulates the idea that in order to reverse the Freudian formula, one has to produce the unconscious: To do so effectively results in a truly revolutionary movement that is born of and thrives on experimentation. According to Deleuze, the unconscious is not something that one has or reaches. It is not a pre-existent through which the subject travels. Desire generates the unconscious while it activates fluxes made of a-signifying signs; signs detached from pre-formulated concepts that keep individuals and social-political fields static. This desire that comes to be while it constructs an experimental deterritorialized space is revolutionary because it always wants

to initiate more relations, connections, agencements and mutations. On the contrary philosophies, arts and sciences that prescribe to psychoanalytic theories ignore, efface and negate such connections and agencements because *they hate real desire and politics.* (Dialogues, 96, 97)

8. Jean Toomer as stated, believed himself to be of "the new American race." This "new race" not only transcends categories imposed by color but by national borders as well. "What I really mean of the American race is the human race. . . . After having for years been hypnotized by labels and suggestions to believe we were less than human, merely Caucasian, or Mongolian, or Negroid, merely African, Russian, Italian, American . . . after having been identified with these surfaces, we are emerging from these limitations, we are waking up . . . we are realizing our basic human stock . . . our fundamental and universal humanity. . . . Those who have or who are approaching this realization- these are the only ones I mean when I say Americans. These Americans are not of America only; they are of the earth. And . . . they of course exist in other national groups. These are the conscious internationalists." (Toomer Reader, 110).

   This ideal "new race" that exceeds the limitations of racial categories and nationalized identities, in part stems from his personal experience being raised in a family that lived on both sides of the "color line." In fact Toomer who lived in the home of his maternal grandfather, who passed for white, was shocked to learn of his black heritage when he was in high school. Jean Toomer knew first hand the arbitrary nature of race distinctions drawn from racial prejudices. Jean Toomer also experienced the potential fluidity of an identity that might be black, white or one fluctuating between the two in a new space, as his grandfather shifted his own racial identification when it served him politically when he ran for U.S. senate. "P.B.S. Pinchback, Jean's grandfather, who for all appearances was a white man, during Reconstruction, had claimed to be black. It was a matter of political expediency, resulting in his election to the U.S. Senate from Louisiana. By the time Jean was born, he left politics, settled well in Washington in a white neighborhood. In high school Jean had the shock of his life when his grandfather told him of his black ancestry. . . . In 1921, when Toomer was taking care of his aged grandparents, who had by then lost their fortune and were living in a black section of Washington." In relation to these experiences Toomer writes, "During this period I read many books on the matter of race and the race problem in America. Rarely had I encountered the nonsense contained in most of these books. It was evident to me, who had seen both the white and the colored worlds, and both from the inside, that the authors of these writings had little or no experience of the matters they were dealing with. Their pages showed very little more than strings of words expressive of personal prejudices and preferences. . . . I wrote a poem called, "The First American," the idea of which was, that here in America we are in the process of forming a new race, that I was one of the first conscious members of this race." (Invisible Darkness, 12)

9. Black writers such as Ralph Ellison and Chester Himes, substantiate Faulkner's ability to "betray his race," and create minoritarian-becomings. For example in an interview with Michael Bandler published originally in the 'Washington Post Book World' in 1972, Chester Himes states that Faulkner is the "blackest" white writer, a writer who knows how to "capture the mentality of blacks." The Question addressed is: "Who's the blackest white writer you know? Himes responds, "You mean the one who understands the black people best? Well, I always felt that Faulkner was able to capture the mentality of blacks in the South better than any other white writer I've read. He exaggerates sometimes but he has some very sharp portraits of black people in the south in his works." (Conversations with Chester Himes, 101)
10. In *Story and Situation* Ross Chamber distinguishes a shift in narrative techniques, one that moves from "narrative" to "narratorial" authority, wherein the writer's aim is no longer to directly transmit information, but to rather arouse "interest," or desire in the reader. Chambers refers to Roland Barthes distinction between the "readerly" verses "writerly" text to better articulate this difference. According to Roland Barthes the *readerly* text, that which is easily "readable" becomes the "writerly" text, (*le scriptible*), when the writer liberates himself from his dependency of the "market place forces that make writing a form of labor and the literary work a commodity." (Story, 13). The "writerly" text becomes multi-layered and provides no one readily discernable meaning. Rather it plays as an autonomous producer of meaning, as a purely verbal construct that inherently mocks any notion of re-presenting a *real* world. It divorces itself from playing a social-political role; as a lens to reflect and support moral values instituted by established social-political codes, laws and order. Barthes states that what is at stake in the writerly text, is to « faire du lecteur, non plus un consommateur, mais un producteur du texte. » (*make the reader no longer a consumer, but a producer of the text*). (Story, 14).

    Ross Chambers points out that even in the nineteenth century; many writers were marginalized as poets, or artists, in an age of bourgeois dominance. Certain artists already had begun to exploit the "modern" literary practice of seducing the reader, producing a style that led to a plurality of possible meanings and interpretive readings. Chambers states that such seduction, which produces authority with an absence of power, converts "historical or social weakness into discursive strength."(Story, 212). This narrative strategy delivered out of a position of weakness subversively asserts alienated, truths, realities, and persons. (Story, 212). In relation to this distinction one must see Anderson's text as falling into the "pre-modern," "readerly" variety.

11. In this passage Deleuze speaks to the power of "chromatic" or "secret" languages, that *'places the public language's system of variables in a state of variation,'* these 'minor' languages ultimately 'grant pragmatism its intensities and values.' These languages that extend "tensors" (pragmatic values) that cause shifts in the major language, and are essential to "agencements," those molecular heterogeneous "block"- movements that stimulate "social and political becomings, ultimately, underlie the success of minoritarian-becomings. Kafka, a Czechoslovakian Jew writing in German, invents a minor language, negotiating all of the variables both to constrict the constants and to expand the variables to make language stammer, or "wail," to draw cries, shouts, pitches, durations, timbres, accents, and intensities from it. Two conjoined tendencies in so-called minor languages have often been noted: an impoverishment, a shedding of syntactical and lexical forms; but simultaneously a strange proliferation of shifting effects . . . we see a rejection of reference points, a dissolution of constant form in favor of differences of dynamism. The closer a language gets to this state, the closer it comes not only to a system of musical notation, but also to music itself. (Thousand, 104, 97)

Bibliography

Allen, Beverly. "Pasolini- The Shadow of his Style." *Stanford Italian Review.* no. 213. March, 1983.

"Alphabetical List of 20th Century Avant-Garde Periodicals in the Sheridan Libraries Collections." <http://www.library.jhu.edu/rsd/other/surreal/alpha.html> March 1, 2003.

Anderson, Sherwood. *Dark Laughter.* New York: Boni and Liveright, 1925.

"Arthur Symons." <http://homepages.nildram.co.uk/~simmers/symchron.htm> April 10, 2002.

Baldwin, James. *Notes of a Native Son.* Boston: Beacon Press, 1955 and 1984.

Ballou, Robert and Murphy, Gardner. *William James on Psychical Research.* New York: The Viking Press, 1960.

Barthes, Roland. *Le degré zéro de l'écriture: suivie de Nouveaux essays critiques.* Paris, Editions de Seuil, 1953 and 1972.

Beaubatie, Yannick, et al. *Tombeau de Gilles Deleuze.* Paris: *Mille Sources,* 2000.

Behler, Ernst. *Confrontations: Derrida, Heidegger, Nietzsche.* Stanford: Stanford University Press, 1991.

Bell, Quentin. *Virginia Woolf: A Biography.* New York: Harcourt Brace, 1974.

"Cinquante ans de philosophie française—Traverses / Professeurs et passeurs" <http://www.adpf.asso.fr/adpf-publi/folio/philo3/04.html> February 2, 2002.

Chambers, Ross. *Story and Situation: Narrative Seduction and the Power of Fiction.* Minneapolis: University of Minnesota Press, 1984.

Colebrook, Claire. *Gilles Deleuze.* London: *Routledge,* 2002.

"The Contribution of American Pragmatism to Contemporary Literary and Cultural Studies." <http://fb14.uni-mainz.de/projects/amst/forum/kinas1.html> 10 May 2002.

"The Crying of Water." <http://www.recmusic.org/lieder/v/various/crying.html> April 2, 2002.

Deleuze, Gilles. *L'Abécédaire de Gilles Deleuze,* avec Claire Parnet <http://www.langlab.wayne.edu/CStivale/D-G/ABC1.html> 4 April, 2002.

Deleuze, Gilles and Claire Parnet. *Dialogues.* Paris: Flammarion, 1996.

Deleuze, Gilles and Guattari, Félix. *Anti-Oedipus: Capitalism and Schizophrenia.* Minneapolis: University of Minnesota Press, 1996.

Deleuze, Gilles and Guattari, Félix. *Kafka: Pour une littérature mineure.* Paris: Les Editions de Minuit, 1975.

Deleuze, Gilles and Guattari, Félix. *Mille Plateaux: Capitalisme et Schizophrénie 2.* Paris: Les Editions de Minuit, 1980.

———. *A Thousand Plateaus: Capitalism and Schizophrenia.* Minneapolis: University of Minnesota Press, 1987.

Deleuze, Gilles. "I have nothing to admit," appeared in *Semiotext(e)* 2.3, 1977, pp. 111–116, The *Anti-Oedipus* Issue. August 1 02. <http:// bush.cs.tamu.edu/~erich/ misc/nothing_to_admit> 20 April, 2001.

Deleuze, Gilles, Lapoujade, David, ed. *L'Ile Désertes et Autres Textes.* Paris: Editions de Minuit, 2002.

———. *Critique et Clinique.* Paris: Les Editions de Minuit, 1993.

———. *Différence et Répétition.* Paris: Presses Universitaires de France, 1968/1981.

———. *Empirisme et Subjectivité: Essai Sur la Nature Humaine Selon Hume.* Paris: Presses Universitaires de France, 1998.

———. *Essays Critcal and Clinical.* Minneapolis: University of Minnesota Press, 1997.

———. *Logique du sens.* Paris: Les Editions de Minuit, 1969.

———. *The Logic of Sense.* New York: Columbia University Press, 1990.

———. "Postscript on the Societies of Control." <http://www.geocities.com/ Athens/Oracle/8810/deleuze.html> 10 April, 2002.

———. "Qu'est-ce que l'acte de création ?" Conférence donnée dans le cadre les mardis de la fondation Fémis. <http://www.webdeleuze.com/html/sommaire.html> 5 March, 2001.

———. *Nietzsche.* Paris: Presses Universitaires de France, 1992.

———. *Nietzsche and Philosophy.* New York: Columbia University Press, 1983.

———. *Nietzsche et La Philosophie.* Paris: Presses Universitaires de France, 1962.

———. *Pourparlers.* Paris: Editions de Minuit, 1990.

Deleuze, Gilles. *Proust et les Signes.* Paris: Presses Universitaires de France,1970

———. *Théorie des Multiplicités chez Bergson.* <http://www.webdeleuze.com/html/ TXT/bergson.html> 19 February, 2002.

De Maupassant, Guy. *Pierre et Jean (Preface).* <http://www.ac-orleans-tours.fr/lettres/ coin_prof/realisme/preface_de_pierre_et_jean.html> 10 July, 2002.

Derrida, Jacques. "I Will Have to Wander All Alone" <http://www.usc.edu/dept/complit/tympanum/1/derrida1.html> 10 June, 2001.

Douglas, Ann. *Terrible Honesty: Mongrel Manhattan in the 1920's.* New York: The Noonday Press, 1996.

Du Bois, W.E.B. "Book Reviews." *The Independent,* November 17, 1904; 57:1152.

———. *The Souls of Black Folk.* New York: Penguin Books, 1989.

———. *Writings.* New York: The Library of America, 1986.

Ellison, Ralph. *Going to the Territory.* New York: Vintage Books, 1986.

———. *Shadow and Act.* New York: Vintage Books, 1995.

Fabre, Michel. *From Harlem to Paris: Black American Writers in France, 1840–1980.* Chicago: University of Illinois Press, 1991.

Fabre, Michel and Robert E. Skinner. *Conversations with Chester Himes.* Mississippi: 1995.

Faulkner, William. *The Sound and the Fury.* Harmondsworth, England: Penguin Modern Classics, 1965.

Fondation Royaumont. *Nietzsche: Cahiers de Royaumont Philosophie VI. (VIIe Colloque- juillet 1964).* Paris: Les Editions de Minuit, 1967.

Gates, Henry Louis Junior. *Figures in Black: Words, Signs, and the "Racial Self."* New York: Oxford University Press, 1987.

Gilroy, Paul. *The Black Atlantic: Modernity and Double Consciousness.* Cambridge: Harvard University Press, 1993.

Goldthorpe, Rhiannon. *La Nausée.* London: Harper Collins Academic, 1991.

Gregory, Horace. *The Portable Sherwood Anderson.* New York: Viking Press, 1977.

Hollier, Denis. *Le Collège de Sociologie (1937–1939).* Paris: Gallimard, 1979 and 1995.

———. *The Politics of Prose: Essay on Sartre.* Minneapolis: University of Minnesota Press, 1986.

James, William. *Essays in Radical Empiricism.* Lincoln: The University of Nebraska Press. 1996.

———. *The Meaning of Truth.* New York: Longmans, Green, and Co. 1909.

———. *A Pluralistic Universe.* Lincoln: The University of Nebraska Press. 1996.

———. *A Pluralistic Universe.* New York: Longmans, Green, and Co. 1916.

———. *Pragmatism.* Toronto and London: Dover Publications. 1995.

———. *Radical Empiricism.* Class Reader/ Engl. 505. The University of Washington. Spring 1996.

———. *Some Problems of Philosophy.* New York: Longmans, Green, and Co. 1911.

———. *Will to Believe.* New York: Longmans, Green, and Co. 1931.

Jarczyk, Gwendoline, and Pierre-Jean Labarrière. *De Kojeve à Hegel: Cent cinquante ans de pensée hégélienne en France.* Paris: Editions Albin Michel, 1996.

Jones, LeRoi. *Blues People: Negro Music in White America.* New York: William Morrow and Company. 1963.

Jones, Randye L. "Biography: H. T. Burleigh (1866–1949). <http://www.afrovoices.com/burleigh.html>

Kayser, Paulette. *Dire du corps -Corporéité et affectivité dans les écrits d'Emmanuel Lévinas.* <http://www.philagora.net/philo/corps1.htm> 20 April, 2003.

Kirkland, Avon. "American Masters: Ralph Ellison." <http://www.pbs.org/wnet/americanmasters/database/ellison_r.html> June, 2002.

Lapoujade, David. *William James Empirisme et Pragmatisme.* Paris: PUF, 1997.

———. "From Transcendental Empiricism to Worker Nomadism: William James." <http://www.warwick.ac.uk/philosophy/pli_journal/pdfs/lapoujade_pli_9.pdf> May 1, 2003.

Larson, Charles. *Invisible Darkness: Jean Toomer and Nella Larsen.* Iowa City: University of Iowa Press, 1993.

"L'effet Deleuze: Philosophie, esthétique, politique." *Magazine Littéraire.* February, 2002: 17–58.

Le Rider, Jacques. *Nietzsche en France: De la fin du XIX siècle au temps présent.* Paris: Presses Universitaires de France, 1999.

Maggiori, Robert. "Deleuze-Guattari: Nous deux." <http://www.liberation.com/livres/25ans/deleuze.html> 17 February, 2003.

Martin, Jean-Clef. *Variations: La Philosophie de Gilles Deleuze.* Paris: Editions Payot & Rivages, 1993.

Mathy, Jean-Philippe. *Extreme Occident: French Intellectuals and America.* Chicago: The University of Chicago Press, 1993.

McNeillie, Andrew, ed. *The Second Common Reader: Virginia Woolf.* New York: Harcourt Brace and Company, 1986.

Miles, Adrian. "The Deleuzeguattarionary" <http://cs.art.rmit.edu.au/deleuzeguattarionary/> January 10, 2002.

Miner, Ward and Thelma M. Smith. *Transatlantic Migration: The Contemporary American Novel in France.* Durham: Duke University Press, 1955.

Mohrt, Michel. *Album William Faulkner.* Paris: Gallimard, Bibliothèque de la Pléiade, 1995.

Morel, Georges. *Nietzsche Analyse de la Maladie.* Paris: Aubier-Montaigne,1971.

———. *Nietzsche Création et Metamorphoses.* Paris: Aubier-Montaigne, 1971.

Murray, Albert. *Stomping the Blues.* New York: Da Capo Press, 1976.

Nielsen, Aldon Lynn. *Black Chant: Languages of African-American Postmodernism.* Cambridge: Cambridge University Press, 1997.

Nietzsche, Friedrich. *The Birth of Tragedy and The Case of Wagner.* New York: Vintage Books, 1967.

———. *Beyond Good and Evil.* Translated by Walter Kaufman. New York: Vintage Books,1989.

———. *Friedrich Nietzsche on Rhetoric and Language.* "On Truth and Lying in an Extra-Moral Sense." New York: Oxford University Press, 1989.

———. *The Gay Science.* Translated by Walter Kaufman. New York: Vintage Books, 1974.

———. *On the Genealogy of Morals and Ecce Homo.* Translated by Walter Kaufman. New York: Vintage Books, 1989.

Ognois, Claude. "Philosophie Pataphysique 3Bis." <http://perso.wanadoo.fr/claude.ognois/ubu3bis.htm> 3 January, 2001.

Pinhas, Richard. "Atmosphère de Paris, septembre 1995." <http://www.webdeleuze.com/html/TXT/rpinhas2.html> 19 April, 2003.

———. Le Plan. <http://membres.lycos.fr/heldon/fr/plan.htm> 10 May, 2001.

———. "WEB DELEUZE: French site with unpublished lectures by Deleuze in French and in English translations." http://www.webdeleuze.com/ 2 July 2002.

Porter, Horace. *Stealing the Fire: The Art and Protest of James Baldwin.* Middletown: Wesleyan University Press, 1989.
Posnock, Ross. *Color and Culture: Black Writers and the Making of the Modern Intellectual.* Cambridge: Harvard University Press, 1998.
Proust, Marcel. *La Prisonnière.* Paris: Pléiade Tome III, 1923.
Sartre, Jean-Paul. *Existentialism and Human Emotions.* New York: Philosophical Library, 1957.
———. *Literary and Philosophical Essays*: New York: Collier Books, 1962.
———. *La Nausée.* Paris: Editions Gallimard, 1938
———. *Situations, II.* Paris: Gallimard, 1948.
Sartre, Jean-Paul and Senghor, Léopold Sedar. *Anthologie de la Nouvelle Poésie Nègre et Malgache de la Langue Française* et *Orphée Noir.*
Silverman, Hugh, et al. Continental Philosophy VII: Philosophy and Desire. New York: Routledge, 2000.
Stein, Gertrude. *The Autobiography of Alice B. Toklas.* New York: Vintage Books, 1990.
———. *Everybody's Autobiography.* New York: Copper Square Publishers, 1971.
———. *Three Lives.* New York: Penguin Books, 1990.
Stearns, Marshall. *The Story of Jazz.* New York: Oxford University Press, 1956 and 1970.
Stivale, Charles. *The Two-Fold Thought of Deleuze and Guattari.* New York: The Guilford Press, 1998.
———. "L'Abécédaire de Gilles Deleuze, avec Claire Parnet Directed by Pierre-Andre Boutang, 1996." 3 May 00 <http:www.langlab.wayne.edu/romance/FreD_G/ABC1.html> November 16, 2002.
Taylor, Alan. "Deleuze and Guattari on the Web." <http://www.uta.edu/english/apt/d%26g/d%26gweb.html> 10 June, 2002.
Thurman, Wallace. *The Blacker the Berry.* New York: Collier Books, 1970.
Toomer, Jean. *Cane.* New York: Liveright, 1975.
Torgovnick, Marianna. *Gone Primative.* Chicago: The University of Chicago Press, 1990.
Wahl, Jean. *Ecrivains et Poètes des Etats-Unis d'Amérique.* Paris: Fontaine,1945.
———. *Le Malheur de la Conscience Dans la Philosophie de Hegel.* Paris: Presses Universitaires de France, 1951.
———. *Petite Histoire de l'Existentialisme.* Paris: Editions Club Maintenant, 1947.
———. *Les Philosophe de L'Existence.* Paris: Collection Armand Colin, 1959.
———. *Les Philosophes Pluralistes d'Angleterre et d'Amérique.* Paris: Librairie Felix Alcan, 1920.
———. *The Philosopher's Way.* New York: Oxford University Press, 1948.
———. *A Short History of Existentialism.* Paris: Editions Club Maintenant, 1949.
———. *Vers le Concret.* Paris: Librairie Philosophique J.Vrin, 1932.
———. *Voices in the Dark: Fifteen Poems of the Prison Camp.* Kirkwood: The Printery, 1974.

Weiss, Lynn. *Gertrude Stein and Richard Wright.* University Press of Mississippi, 1998.

Werner, Craig Hansen. *Playing the Changes: From Afro-Modernism to the Jazz Impulse.* Chicago: The University of Illinois Press,1994.

Wright, Richard. *Uncle Tom's Children.* New York: Harper Perennial, 1993.

Index

L

M

N

P

For Product Safety Concerns and Information please contact our EU
representative GPSR@taylorandfrancis.com
Taylor & Francis Verlag GmbH, Kaufingerstraße 24, 80331 München, Germany

www.ingramcontent.com/pod-product-compliance
Ingram Content Group UK Ltd.
Pitfield, Milton Keynes, MK11 3LW, UK
UKHW022202190926
475702UK00003B/162
* 9 7 8 0 4 1 5 9 7 5 3 5 3 *